Chemical Sensors

CW00546496

Robert W. Cattrall

Centre for Scientific Instrumentation
School of Chemistry
La Trobe University
Melbourne
Australia

Series sponsor: **ZENECA**

ZENECA is a major international company active in four main areas of business: Pharmaceuticals, Agrochemicals and Seeds, Specialty Chemicals, and Biological Products.

ZENECA's skill and innovative ideas in organic chemistry and bioscience create products and services which improve the world's health, nutrition, environment, and quality of life. ZENECA is committed to the support of education in chemistry and chemical engineering.

OXFORD   NEW YORK   MELBOURNE
OXFORD UNIVERSITY PRESS
1997

Oxford University Press, Great Clarendon Street, Oxford OX2 6DP

Oxford   New York
Athens   Auckland   Bangkok   Bogota   Bombay   Buenos Aires
Calcutta   Cape Town   Dar es Salaam   Delhi   Florence   Hong Kong
Istanbul   Karachi   Kuala Lumpur   Madras   Madrid   Melbourne
Mexico City   Nairobi   Paris   Singapore   Taipei   Tokyo   Toronto   Warsaw
and associated companies in
Berlin   Ibadan

Oxford is a trade mark of Oxford University Press

Published in the United States
by Oxford University Press Inc., New York

A catalogue record for this book is available from the British Library

Library of Congress Cataloging in Publication Data
(Data available)

ISBN 0 19 850090 4

Typeset by EXPO Holdings, Malaysia

Printed in Great Britain by
The Bath Press, Avon

Series Editor's Foreword

Oxford Chemistry Primers are designed to provide clear and concise introductions to a wide range of topics that may be encountered by chemistry students as they progress from the freshman stage through to graduation. The Physical Chemistry series will contain books easily recognized as relating to established fundamental core material that all chemists will need to know, as well as books reflecting new directions and research trends in the subject, thereby anticipating (and perhaps encouraging) the evolution of modern undergraduate courses.

In this Physical Chemistry Primer, Bob Cattrall has produced a stimulating, easy-to-read introduction to the use of *Chemical Sensors*. The Primer will interest all students (and their mentors) who wish to appreciate the principles and applications of this increasingly important topic.

Richard G. Compton
Physical and Theoretical Chemistry Laboratory, University of Oxford

Preface

The aim of this primer is to introduce senior undergraduate students to the exciting world of chemical sensors. An attempt has been made to write it in an easy to understand form so that it can be used by students in all the relevant branches of science. Emphasis is placed on describing the scientific principles upon which the various sensor types are based, as well as indicating some of their applications. The primer should be equally valuable to chemistry students and students in the biological and medical sciences, as well as being of interest to electronic engineering students as they explore the interface between chemistry and electronics. But most of all, I hope it will inspire our future young researchers and lead them into a world which will be dominated by our need and desire to monitor in real time everything around us. The future is an exciting one and we will see the next generation of chemical sensors integrated into miniaturized total analysis systems (μTAS) which will be applied in nearly every aspect of our lives. This will require the knowledge of the scientist as well as the skill and ingenuity of the engineer.

I owe special thanks to Professor Henry Freiser who led me into the fascinating world of chemical sensors, and to my wife Bev who put up with me retreating into my study on so many occasions to work on the manuscript but continued to offer encouragement.

Melbourne   R.W.C.
February 1997

Contents

1 Monitoring our environment

1.1 Why do we need chemical sensors?

There is an escalating need and desire for us to monitor all aspects of our environment in real time and this has been brought about by our increasing concerns with pollution, our health, and safety. There is also a desire to determine contaminants and analytes at lower and lower levels and one could say that the aim of all modern analytical chemistry is to lower the detection limits and to improve the accuracy and precision at those limits. Instrumentation in analytical chemistry has become so sophisticated that we are now able to detect chemicals in amounts smaller than we ever imagined of a few years ago. In fact, this has shown that manufactured chemicals and by-products have been introduced into almost every aspect of our environment and lives.

Because of this desire and need to monitor everything around us there is a tremendous input of energy and resources into developing sensors for a multitude of applications. The end result of all this research will one day provide us with portable, miniature, and intelligent sensing devices to monitor almost anything we wish. For example, take our health, one can imagine in the future having a credit card size self-diagnostic unit with a multitude of chemical sensors and biosensors built into it so that we can monitor our well-being at any instant. A person may be feeling unwell and so licks the sensing surface of their diagnostic unit. Immediately, the liquid crystal display flashes up the message, 'you have an influenza virus, take aspirin and rest'.

In the future, sensors will be part of almost everything to do with our lives.

In monitoring the environment one can imagine similar devices which could be used to test for say heavy metal pollutants in natural waters or the presence of bacteria in drinking water, swimming pools, or at beaches. Bathers might carry such a device with them to test the water before swimming. The possibilities are limitless and are controlled only by one's imagination. Well, that is not strictly true, the possibilities are really controlled by the chemistry, physics, and electronics of such devices and the state of the art at that particular point in time. We should never lose sight of the fact that all these sensing systems depend on sound basic science. Sensor research has brought together a unique team of chemists, biologists, physicists, and electronic engineers and is thus a truly interdisciplinary field.

Sensors can be categorized into two general groups. There are physical sensors which are sensitive to such physical responses as temperature, pressure, magnetic field, and force and these do not have a chemical interface. Then, there are chemical sensors which rely on a particular chemical reaction for their response. It is this latter group which is discussed in this primer.

1.2 What is a chemical sensor?

A chemical sensor is a device which responds to a particular analyte in a selective way through a chemical reaction and can be used for the qualitative or quantitative determination of the analyte.'

The term 'analyte' is used in analytical chemistry to define the species being analysed in a sample. An analyte can be an atom, ion, or molecule in a solution, a solid, or in the gas phase.

It can be seen that such a definition encompasses all sensors based on chemical reactions including biosensors which make use of highly specific and sensitive biochemical and biological reactions for species recognition.

There are two parts to a chemical sensor. Firstly, there is the region where the selective chemistry takes place and then there is the transducer. The chemical reaction produces a signal such as a colour change, the emission of fluorescent light, a change in the electrical potential at a surface, a flow of electrons, the production of heat, or a change in the oscillator frequency of a crystal. The transducer responds to this signal and translates the magnitude of the signal into a measure of the amount of the analyte.

1.3   Types of chemical sensors

In this primer, chemical sensors are categorized into the following groups according to the transducer type.

1. *Electrochemical.* These include potentiometric sensors (ion-selective electrodes, ion-selective field effect transistors (ISFETs and CHEMFETs)) and voltammetric/amperometric sensors including solid electrolyte gas sensors. Semiconductor-based gas sensors which rely on the measurement of the electrical conductivity of the semiconductor could also be included in this category, although there is no electrochemical reaction taking place in the response mechanism.

2. *Optical.* In optical sensors there is a spectroscopic measurement associated with the chemical reaction. Optical sensors are often referred to as 'optodes' and the use of optical fibres is a common feature. Many biosensors make use of optical measurements. Absorbance, reflectance, and luminescence measurements are used in the different types of optical sensors.

3. *Mass sensitive.* These make use of the piezoelectric effect and include devices such as the surface acoustic wave (SAW) sensor and are particularly useful as gas sensors. They rely on a change in mass on the surface of an oscillating crystal which shifts the frequency of oscillation. The extent of the frequency shift is a measure of the amount of material adsorbed on the surface.

4. *Heat sensitive.* These are often called calorimetric sensors in which the heat of a chemical reaction involving the analyte is monitored with a transducer such as a thermistor or a platinum thermometer. Flammable gas sensors make use of this principle.

These sensor types are described in the various chapters of the primer. As mentioned before, biosensors are chemical sensors and are discussed along with all the other sensors and are not treated in a separate chapter, however, they make up such an important class of sensors that a few additional comments need to be made.

1.4   Biosensors

Biosensors make use of the fundamental chemical reactions which life itself depends on. Antibody/antigen, enzyme/substrate, and receptor/hormone reactions have all been utilized to yield highly selective and sensitive sensors

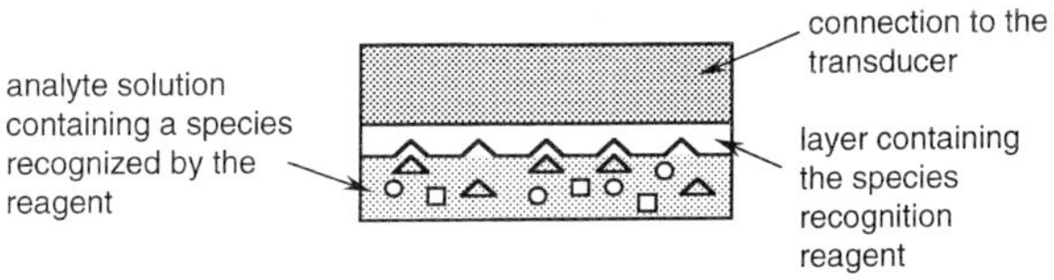

Fig. 1.1 A biosensor.

for particular analyte species. A biosensor is often illustrated in a diagram similar to that shown in Fig. 1.1, however, such a representation could apply to all the other chemical sensor types in which a reagent shows an affinity for a particular species. The 'lock and key' representation is designed to illustrate the highly selective reactions which occur between biological molecules.

In a biosensor the species recognition reagent is often a macromolecule which is immobilized into a membrane or chemically bound to a surface in contact with the analyte solution. A specific chemical reaction then takes place between the reagent and the analyte. This can be the direct binding of the reagent with the analyte as occurs in antibody/antigen reactions or the catalytic interaction of an immobilized enzyme with the analyte to produce products which can be detected. The transducer types used with biosensors include all those listed above.

The immune system in animals is an example of the development of highly specific molecules for species recognition and this has been utilized to produce extremely selective and sensitive test procedures in clinical biochemistry such as radio immuno assay (RIA) and enzyme-linked immunosorbent assay (ELISA) methods. Considerable work has been undertaken to produce immuno-sensors based on these principles. Antibodies against almost any substance which can interact with a biological system, e.g. drugs, toxins, metabolites, and pesticides, can be produced and so this is an area of great interest and activity for the development of sensors.

An interesting and important development comes from synthetic organic chemistry where some very creative and elegant research has produced organic molecules containing specific binding sites which mimic the active sites in naturally occurring biological molecules, and very selective and sensitive sensors have been made with these compounds. Many advances in this area can be expected in the future.

In many instances, there is already a well-developed technology in clinical medicine (and to a lesser extent in environmental chemistry) based on the 'test strip' approach in which single-use, reagent-impregnated paper strips are used to perform rapid tests on biological fluids for a number of biochemical functions. These are inexpensive and useful tests which can be carried out by physicians or patients. One example is the test for glucose in blood or urine and the question has been raised as to why scientists are investing enormous amounts of resources in a quest for glucose biosensors. The answer lies in the ultimate desire to have implantable sensors for the continuous monitoring of a patient in real time and to have the output from the sensor interfaced to a feedback system to deliver the required dose of medication when it is needed. This, of course, relates to our desire to continuously monitor everything about our environment (and to take corrective action) and is the main driving force behind sensor research.

2 Solid state potentiometric chemical sensors

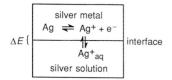

Fig. 2.1 Establishment of the interfacial potential for silver metal.

The picture represented in this primer of the interface between two phases is a simplified one designed to enable even students with only a basic knowledge of chemistry to see how interfacial electrical potentials arise. For a thorough understanding of the theory of interfacial potentials and the 'electrical double layer' students are directed to the text *Modern electrochemistry*, Vol. 2, J. O'M. Bockris and A. K. N. Reddy, Plenum Press, NY, 1970.

Ohm's Law is $V = i.R$, where 'V' is the EMF in volts, 'i' is the current in amperes, and 'R' is the resistance in ohms.

Originally, these electrodes were termed 'specific-ion electrodes', then the term 'ion-selective electrode' was used and now we often refer to them as 'potentiometric chemical sensors'.

2.1 The technique of potentiometry

Potentiometric chemical sensors make use of the development of an electrical potential at the surface of a solid material when it is placed in a solution containing ions which can exchange with the surface. The magnitude of the potential is related to the number of ions in the solution. Perhaps the simplest case to use as an example is that of a piece of silver wire which is placed in a solution containing silver ions (Fig. 2.1). The silver metal consists of a periodic network of positive ions and a pool of mobile electrons. A few Ag^+ ions at the surface distribute into the aqueous solution in the region of the surface and become hydrated (Ag^+_{aq}) which leaves an equivalent number of free electrons on the metal side of the solution/metal interface. In other words, there has been a charge separation across the interface which gives rise to an electrical potential difference.

The magnitude of the potential difference will depend on how many Ag^+ ions distribute across the interface and this is related to the concentration (actually to the thermodynamic activity) of silver in the solution. It is not possible to measure the interfacial potential directly but it can be measured in a suitable electrochemical cell. This is the technique of potentiometry and it is said that the measurement of the cell potential is made under a zero current condition. In practice of course, such a condition is not attainable since the very nature of the potential measurement means that a very small current must flow. The millivoltmeter which is used to measure the potential operates according to Ohm's Law but the current flow is in the μA range and does not normally disturb the equilibrium potential at the surface. Thus, the assumption that the measurement of the potential is made under an essentially zero current condition is reasonably correct.

2.2 Ion-selective electrodes

Ion-selective electrodes are classified as potentiometric chemical sensors since some selective chemistry takes place at the surface of the electrode producing an interfacial potential. Species recognition is achieved with a potentiometric chemical sensor through a chemical equilibrium reaction at the sensor surface. Thus, the surface must contain a component which will react chemically and reversibly with the analyte. This is achieved by using ion-selective membranes which make up the sensor surface. In the early literature, ion-selective electrodes which use such membranes, were often referred to as being 'specific' for a particular ion. We now know that other ions can influence the chemical equilibrium at the sensor surface and so it is more correct to use the term 'selective' rather than 'specific' to describe these sensors.

2.3 Sensor membranes

There are four categories of membranes used in potentiometric chemical sensors.

1. Glass membranes. These are selective for ions such as H^+, Na^+, and NH_4^+.

2. Sparingly soluble inorganic salt membranes. This type consists of a section of a single crystal of an inorganic salt such as LaF_3 or a pressed powdered disc of an inorganic salt or mixtures of salts such as $Ag_2S/AgCl$. Such membranes are selective for ions such as F^-, S^{2-}, and Cl^-.

3. Polymer-immobilized ionophore membranes. In these, an ion-selective complexing agent or ion-exchanger is immobilized in a plastic matrix such as poly(vinyl chloride).

4. Gel-immobilized and chemically bonded enzyme membranes. These membranes use the highly specific reactions catalysed by enzymes. The enzyme is incorporated into a matrix or bonded onto a solid substrate surface.

The first two of these categories are often classed as 'solid state' membranes and are discussed in this chapter. The other membrane categories are discussed in the next chapter of this primer.

2.4 Glass membrane sensors

The common glass electrode for pH measurement is an example of a potentiometric sensor and has been known for more than 80 years, well before the development of the so-called new breed of ion-selective electrodes such as the fluoride electrode in the 1960s. The membrane in a pH-electrode is essentially a sodium silicate glass made by fusing a mixture of Al_2O_3, Na_2O, and SiO_2. This is then blown into a membrane which is 0.3–0.4 mm in thickness (hence the fragility of a glass pH-electrode). With a low amount of Al_2O_3, a glass with a predominant pH response is obtained. Increasing the amount of Al_2O_3 in the glass leads to an increasing response to other monovalent cations such as Na^+, Li^+, and K^+. In all cases, however, the glass membrane also responds to pH.

For a more detailed description of the glass electrode, see *Glass electrodes for hydrogen and other cations. Principles and practice*, (ed., G. Eisenman) Marcel Dekker, NY, 1967.

Glass membranes have a very high electrical resistance in the $M\Omega$ range, however, they must conduct ionic charge to some extent in order to be able to make measurements with them. There is some evidence that sodium ions are the charge carriers.

$1\,M\Omega = 10^6\,\Omega$

All potentiometric chemical sensors need conditioning before use and the pH-sensitive glass electrode is no exception. The conditioning process for the glass membrane is usually carried out by soaking it in 0.1 M HCl for an hour or two. The glass structure can be thought of as consisting of fixed anionic silicate sites ($-SiO_4^{4-}$) which bind monovalent cations reversibly. Thus, in the case of the pH-sensitive membrane, the conditioning process involves the formation of silanol groups ($-SiO^-H^+$) on the glass surface according to eqn 2.1. It is also thought that the conditioning process creates a hydrated gel layer on the glass surface (10^{-5}–10^{-4} mm thick) which is necessary for the membrane to function.

Silicate structures are dominated by tetrahedral SiO_4^{4-} units linked together by sharing oxygen atoms. There are terminal $-SiO^-$ groups which are involved in exchange reactions with cations.

$$-SiO^-Na^+ + H^+ + \rightleftharpoons -SiO^-H^+ + Na^+ \qquad (2.1)$$

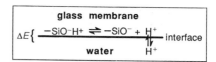

Fig. 2.2   Establishment of the interfacial potential for the glass membrane.

The $-SiO^-H^+$ group is reversible to H^+ exchange in preference to other monovalent cations and so species recognition occurs at the fixed $-SiO^-$ site. A change in the glass structure, as occurs by adding Al_2O_3, makes it easier for other monovalent cations to occupy the $-SiO^-$ site and an increase in selectivity for these ions is seen.

When a glass membrane is put into water, a charge separation process occurs across the glass/water interface giving rise to an electrical potential difference. A qualitative picture can be constructed for the mechanism by which the interfacial potential is established as represented in Fig. 2.2.

The silanol groups ($-SiO^-H^+$) dissociate according to the equilibrium shown and the hydrogen ions produced become hydrated and distribute across the interface to the water side. This produces a separation of charges with the fixed negatively charged silicate sites remaining on the glass membrane side of the interface. Thus, there is an electrical potential difference across the glass/water interface and the magnitude of this depends on the position of the equilibrium which, in turn, depends on the number of hydrogen ions (actually the thermodynamic hydrogen ion activity) in the aqueous solution originally (i.e. on the pH, since $pH = -\log_{10} a_{H^+}$). The higher the activity of hydrogen ions in the sample solution the lower is the negative charge on the membrane side of the interface. The same is true about the inside surface of the glass membrane, but the pH of the internal filling solution is kept constant and so the inside surface/solution potential difference remains constant.

Experimentally, the potential at the water/glass interface is measured with a high input impedance millivolt meter (modern meters have a $10^{12}\ \Omega$ input impedance) in an electrochemical cell which also contains a reference electrode. Such a cell is shown in Fig. 2.3.

According to convention, this electrochemical cell is written in the following way. The primary chemical sensor is written on the right hand side with the reference electrode on the left. The single vertical lines represent solid/solution interfaces and the double vertical line represents a junction between two liquid phases.

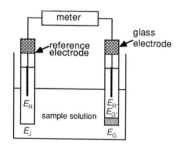

Fig. 2.3   Electrochemical cell.

$$Hg,\ Hg_2Cl_2|satd\ KCl\ \|\ sample\ solution\ |\ glass\ membrane\ |$$
$$0.1\ M\ HCl\ |\ AgCl,\ Ag$$

In Fig. 2.3, the glass and reference electrodes are shown as two separate electrodes, however, combination glass/reference electrodes are common and one is illustrated in Fig. 2.4. Students often confuse this with a single electrode but careful examination will show that there are actually two electrodes built into a single barrel. The reference electrode is often a saturated calomel electrode or a $Ag/AgCl/KCl_{aq}$ electrode. Reference electrodes are simply examples of potentiometric chemical sensors which respond reversibly to the chloride ion (in the old convention, such electrodes were referred to as 'electrodes of the second kind').

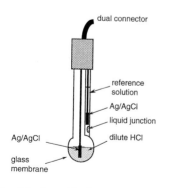

Fig. 2.4   A combination pH/reference electrode.

The potential of the electrochemical cell depicted in Fig. 2.3 is given by the difference between the potential of the right hand side and the left hand side of the cell.

$$E = E_{RHS} - E_{LHS} \tag{2.2}$$

$$\text{or } E = E_G + E_{G'} + E_{R'} - E_R - E_J \tag{2.3}$$

where E_G, $E_{G'}$, $E_{R'}$, E_R, and E_J are the potentials of the two glass/water interfaces (E_G, $E_{G'}$), the two reference element potentials ($E_{R'}$, E_R), and the liquid junction potential (E_J).

It is assumed in making measurements with this cell that all the above interface potentials are constant and that the only one which changes with the solution pH is E_G.

The cell potential is given by the following form of the Nernst equation.

$$E = \text{constant} + RT/nF \ln a_{H^+_{aq}} \tag{2.4}$$

where the constant term contains the interface potentials in the cell which remain constant. In terms of the pH ($-\log_{10} a_{H^+}$), eqn (2.4) is written as,

$$E = \text{constant} - 59.1 \text{ pH} \tag{2.5}$$

Note that the Nernst factor, $2.303RT/nF$, for a monovalent cation is $+59.1$ mV at 25 °C.

Thus, the glass pH-sensitive membrane responds according to eqn 2.5 (i.e. in a Nernstian fashion) over the pH range of about 1 to 14. It should be remembered, however, that at high pH values in the presence of monovalent ions such as K^+ and Na^+ the glass membrane may also respond to these ions producing the so-called 'alkali ion' error.

2.5 Dissolved gas sensing probes

These make use of glass membrane sensors, particularly for pH, to determine the concentration of dissolved gases in aqueous solutions. The term 'probe' is used because the system is a complete electrochemical cell in its own right. The most common of these is the dissolved ammonia probe which is shown in Fig. 2.5.

A combination pH/reference electrode is located inside the barrel of the ammonia probe. A gas permeable membrane (something like a dialysis membrane) is inserted in the end of the barrel and the barrel is filled with an ammonium chloride solution (often 0.1 M). The glass electrode is situated close to the inside surface of the gas permeable membrane and monitors the pH of the ammonium chloride solution. The probe is immersed in the sample which has been treated with a sodium hydroxide solution to produce 'free' ammonia. The ammonia gas diffuses through the gas permeable membrane into the inner ammonium chloride solution and the pH shifts according to the equilibrium shown in eqn 2.6. This is, of course, the same equilibrium reaction which produces the ammonia gas in the first place from the sample in the presence of a strong sodium hydroxide solution (any NH_4^+ is converted to NH_3).

$$NH_3 + H_2O \underset{K_a}{\overset{K_b}{\rightleftharpoons}} NH_4^+ + OH^- \tag{2.6}$$

The liquid junction potential arises from the unequal diffusion of ions across the interface separating two solutions. The size of the liquid junction potential depends on the composition of each of the solutions but is usually a few millivolts. Care needs to be taken to make sure that the reference solution is of similar composition to the sample to minimize the size of the potential. In any event, the liquid junction potential must be constant.

For a better understanding of the Nernst equation and its derivation, the reader is referred to Primer 41 in this series entitled *Electrode potentials* by R. G. Compton and G. H. W. Sanders.

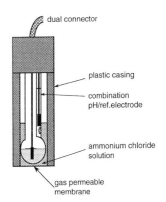

Fig. 2.5 The dissolved ammonia probe.

Poly(tetrafluoroethylene) (PTFE or Teflon®) membranes (plumbers' tape) are often used.

The term 'dissolved ammonia' in waste or natural waters includes both NH_4^+ and NH_3. The amount of each form present will depend on the sample pH.

The ammonia probe responds to concentrations in the range of 10^{-6} M (0.01 mg/L) to 10^{-1} M (1000 mg/L) and is thus ideal for the determination of dissolved ammonia in waste waters and natural waters. It can also be used for the determination of NH_4^+ as shown above and for organic nitrogen following Kjeldahl digestion. The probe is unaffected by coloured or turbid solutions and there are few interferences since these are separated by the gas permeable membrane. Exceptions are small molar mass amines which can diffuse through the membrane and change the pH of the ammonium chloride solution.

Remember, the glass pH-sensitive membrane within the probe responds to the pH of the ammonium chloride solution according to eqn 2.5 and therefore indirectly to the ammonia concentration of the sample as shown in eqn 2.7.

$$E = \text{constant} - 59.1 \log[NH_3] \qquad (2.7)$$

(It should be noted that eqn 2.7 is only valid at a constant ionic strength of standards and samples.) Other types of gas sensing probes together with the sensing membrane used and the equilibrium reaction involved are given in Table 2.1.

The response of gas sensing probes can be slow (several minutes) because of the requirement for the gas to diffuse through the gas permeable membrane and equilibrate with the inner electrolyte solution. However, this is not a great problem at the higher concentrations (95% response in one minute).

2.6   Sparingly soluble inorganic salt membrane sensors

There are a few sparingly soluble inorganic salts which are ionic conductors at room temperature and which can undergo ion-exchange interactions at the solid/water interface to generate an interfacial potential in a similar way to glass membranes. LaF_3, Ag_2S, and to a lesser extent CuS are examples and these can be used in sensor membranes. Three types of sensor membranes employing sparingly soluble inorganic salts are known. These are (a) single crystal membranes, (b) pressed powder membranes, and (c) membranes where the powdered salt is held together by an inert binder (usually a polymer).

Single crystal membranes

A rod consisting of a single crystal of LaF_3 can be grown by melting the powder in an induction furnace under vacuum using the so called 'Bridgman' method and cooling the melt over a period of hours. The rod can then be cut with a diamond saw to produce a thin disc.

In 1966, it was discovered that a slice of a single crystal of lanthanum fluoride attached to the end of an electrode barrel could be used to sense the fluoride ion in aqueous solution. This pioneering work led to the emergence of a new breed of potentiometric chemical sensors and to a field of research which has seen tremendous expansion since that time.

The first report of the fluoride ion-selective electrode was made by M. S. Frant and J. W. Ross in 1966 (*Science*, **154**, 1553).

Table 2.1   Types of gas sensing probes

Gas sensed	Inner solution	Equilibrium reaction	Sensing membrane
CO_2	$NaHCO_3$	$CO_2 + H_2O = HCO_3^- + H^+$	pH–glass
SO_2	$NaHSO_3$	$SO_2 + H_2O = HSO_3^- + H$	pH–glass
HF	H^+	$HF = H^+ + F^-$	$F–LaF_3$
H_2S	pH 5 buffer	$H_2S = HS^- + H^+$	
		$HS^- = S^{2-} + H^+$	$S^{2-}–Ag_2S$
HCN	$KAg(CN)_2$	$Ag(CN)_2^- = Ag^+ + 2CN^-$	$Ag^+–Ag$

LaF$_3$ is an ionic conductor at room temperature and it is the fluoride ion which is the mobile ion in the crystal lattice. Even so, the conductivity of fluoride is low (10^{-7} Ω^{-1}cm^{-1}) and is increased by adding about 1% of europium difluoride to the crystal when it is grown from the melt. LaF$_3$ has a hexagonal crystal structure and may be pictured as layers of LaF$_2^{2+}$ ions with layers of fluoride ions on either side of the LaF$_2^{2+}$ walls. The fluoride ion moves through the crystal lattice by a defect mechanism hopping from one vacancy (Frenkel defect) to another. The addition of EuF$_2$ increases the number of vacancies and hence the conductivity.

A typical configuration for the commercial fluoride ion-selective electrode is shown in Fig. 2.6. The electrochemical cell for making measurements with this electrode is as follows.

$$\text{Hg, Hg}_2\text{Cl}_2 \mid \text{satd KCl} \parallel \text{sample solution} \mid \text{LaF}_3 \mid 0.1 \text{ M NaCl/NaF} \mid \text{AgCl, Ag}$$

The membrane potential is established according to the scheme shown in Fig. 2.7. Fluoride ions will distribute themselves across the crystal/water interface according to the number of fluoride ions initially in the water phase. The fluoride ions which cross to the aqueous side of the interface leave what is essentially a positively charged vacancy in the LaF$_3$ crystal lattice, the number of such vacancies being determined by the number of fluoride ions which leave. In this way, a charge separation occurs across the crystal/water interface giving rise to the interfacial potential, the size of which is dependent on the activity of fluoride in the water phase. The mechanism is similar to that for the glass membrane except that the charges are opposite in sign. The higher the fluoride ion activity in the sample solution the less positive is the charge on the membrane side of the interface.

Obviously, any other ion which can occupy a membrane vacancy will affect the membrane potential but fortunately, in the LaF$_3$ case, the hydroxide ion is the only serious interferent with the fluoride ion-selective electrode. In any event, this is counteracted by buffering the sample solution to pH 5.5. The fluoride ion-selective electrode is used to determine fluoride over the concentration range of 0.1–2000 mg/L. In cities where fluoride is added to the water supply, the electrode is excellent for monitoring the fluoride which is generally maintained at about 1 mg/L.

The cell potential for the electrochemical cell used to make measurements of the fluoride ion activity is given by the Nernst equation.

$$E = \text{constant} - 59.1 \log a_{\text{F}^-_{\text{aq}}} \tag{2.8}$$

Pressed powder membranes

Silver sulfide (Ag$_2$S) is another example of a sparingly soluble inorganic salt which is an ion conductor at room temperature. Also powdered Ag$_2$S has the very useful property of being easily pressed into a mechanically stable membrane such as a thin disc. Thus, it is not necessary to grow a single crystal

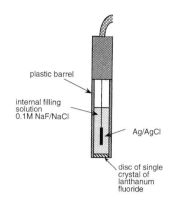

plastic barrel

internal filling
solution
0.1M NaF/NaCl

Ag/AgCl

disc of single
crystal of
lanthanum
fluoride

Fig. 2.6   The fluoride ion-selective electrode.

Analytical chemists prefer to use the terms of mg/L and μg/L to describe the concentration of an analyte rather than the older terms of ppm and ppb. Biologists on the other hand use terms like mM and μM, where M is mol/L.

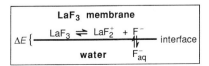

Fig. 2.7   The establishment of the interfacial potential for the LaF$_3$ membrane.

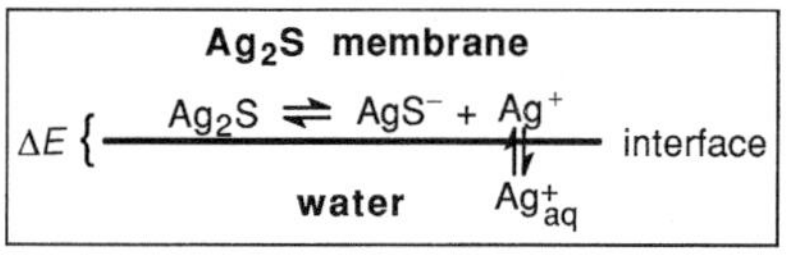

Fig. 2.8   The establishment of the interfacial potential for the Ag$_2$S membrane.

Ag$_2$S can be precipitated from an AgNO$_3$ solution by bubbling H$_2$S gas through the acidified solution. The black powder is collected by filtration, dried, and pressed into a disc in a hydraulic press similar to that used to make KBr discs for infrared spectroscopy. The disc obtained has a shiny metallic lustre.

of Ag$_2$S from a high temperature melt and the powdered salt can simply be precipitated from solution. Powdered LaF$_3$, on the other hand, does not form a mechanically stable disc under pressure and so is best used as a crystal.

An Ag$_2$S membrane in an electrode configuration similar to that shown in Fig. 2.6 for fluoride can be used in the following electrochemical cell to sense both the silver ion and the sulfide ion. This is the system used in the commercial sulfide ion-selective electrode.

Hg, Hg$_2$Cl$_2$ | satd KCl ‖ sample solution | Ag$_2$S | 0.1 M NaCl/Na$_2$S | AgCl, Ag

It is the silver ion which is the mobile ion in the membrane (through a defect mechanism similar to that for fluoride) and sulfide is sensed indirectly through the solubility product term for Ag$_2$S. The membrane potential is again established by a charge separation mechanism in which silver ions distribute across the membrane/water interface as shown in Fig. 2.8. Each silver ion leaving the membrane phase leaves behind what is essentially a negatively charge vacancy in the crystal lattice. The same mechanism can be proposed regardless of whether the Ag$_2$S membrane is a single crystal or a pressed powder disc. The Nernst equation for the Ag$_2$S case is,

$$E = \text{constant} + 59.1 \log a_{Ag^+_{aq}} \tag{2.9}$$

The response of the Ag$_2$S membrane to the sulfide ion can be understood by writing the expression for the solubility product for Ag$_2$S, $K_{sp_{Ag_2S}}$, and including this term in eqn 2.9.

$$\text{solubility product} \qquad K_{sp_{Ag_2S}} = a^2_{Ag^+} \cdot a_{s^{2-}} \tag{2.10}$$

$$\text{rearranging} \qquad a_{Ag^+} = (K_{sp_{Ag_2S}}/a_{s^{2-}})^{\frac{1}{2}} \tag{2.11}$$

$$\text{substitution} \qquad E = \text{constant} + 59.1 \log((K_{sp_{Ag_2S}}/a_{s^{2-}})^{\frac{1}{2}}) \tag{2.12}$$

$$E = \text{constant}' - 29.6 \log a_{S^{2-}} \tag{2.13}$$

It can be seen from the above equations that the sulfide ion activity controls the activity of the silver ion at the membrane surface and so controls the membrane potential, even though it is the silver ion which distributes across the interface and is the mobile ion in the membrane. In other words, the sulfide ion is sensed indirectly.

There are a number of pressed powder type membranes which are mixtures of silver sulfide and other sparingly soluble inorganic salts. These make use of the ion conduction properties of silver sulfide to provide a membrane with the desired electrical properties, but which use the other salt to obtain selectivity for an ion other than silver or sulfide. Membranes are made by pressing discs from the mixed powders and typical mixtures are given in Table 2.2.

It is interesting to note that the commercial cyanide ion-selective electrode

The CN$^-$ ion interferes strongly with the I$^-$ ion response and vice versa.

uses exactly the same membrane as the iodide electrode. In all these cases, the primary response of the membrane is to the silver ion according to eqn 2.9 and

Table 2.2 Mixed powder membranes

Ion sensed	Membrane
Cl^- (or Br^-, I^-)	$Ag_2S/AgCl$ (or $AgBr$, AgI)
CN^-	Ag_2S/AgI
Cd^{2+}	Ag_2S/CdS
Pb^{2+}	Ag_2S/PbS
Cu^{2+}	Ag_2S/CuS

response to the other ions is again achieved via the solubility product of the respective sparingly soluble salt. The solubility product of the salt must be higher than that for silver sulfide. In the case of the Cl^- sensor, the solubility product term for AgCl is incorporated into the Nernst equation as follows.

$$K_{sp_{AgCl}} = a_{Ag^+} \cdot a_{Cl^-} \tag{2.14}$$

$$E = \text{constant} + 59.1 \log(K_{sp_{AgCl}}/a_{Cl^-}) \tag{2.15}$$

$$E = \text{constant}' - 59.1 \log a_{Cl^-} \tag{2.16}$$

In the case of say the Cd^{2+} sensor, the solubility product terms for both Ag_2S and CdS must be included in the Nernst equation to obtain the response equation for Cd^{2+}.

> Both the Ag^+ and the S^{2-} ions must be absent in the sample when determining Cd^{2+}.

$$\text{solubility product terms} \quad K_{sp_{Ag_2S}} = a_{Ag^+}^2 \cdot a_{S^{2-}} \tag{2.17}$$

$$K_{sp_{CdS}} = a_{Cd^{2+}} \cdot a_{S^{2-}} \tag{2.18}$$

solving eqns 2.17 and 2.18 simultaneously

$$a_{Ag^+} = (K_{sp_{Ag_2S}}/K_{sp_{CdS}})^{\frac{1}{2}} \cdot a_{Cd^{2+}}^{\frac{1}{2}} \tag{2.19}$$

inclusion in eqn 2.9

$$E = \text{constant} + 59.1 \log((K_{sp_{Ag_2S}}/K_{sp_{CdS}})^{\frac{1}{2}} \cdot a_{Cd^{2+}}^{\frac{1}{2}}) \tag{2.20}$$

$$\text{simplifying} \quad E = \text{constant}' + 29.6 \log a_{Cd^{2+}} \tag{2.21}$$

Pungor type sensor-membranes

These membranes use powdered sparingly soluble inorganic salts as the electroactive reagents but, instead of pressing the material into a disc, it is formed into a membrane using a polymeric binder. The original Pungor membranes used polysiloxane as the binder and contained about 50% of the electroactive material. There are a number of silicone rubber compositions available commercially and most of these are suitable for use as a binder. A typical membrane can be prepared by mixing the silicone rubber (e.g. General Electric's Clear Seal® or Dow Corning's Silastic Clear Sealer®) with a sufficient amount of finely powdered electroactive material to obtain a coherent slurry which contains about 50% of the material. The slurry is cured in air and forms a mechanically strong sensor-membrane. Membranes can be

> After the distinguished Hungarian chemist, Professor Erno Pungor.

cut to a suitable size and glued to the end of a barrel to make an ion-selective electrode similar to that shown in Fig. 2.6.

Pungor type membranes have been studied for all of the sparingly soluble salts described above and commercial versions have been produced. These membranes have almost identical response characteristics to the pressed disc membranes and the response mechanism is the same.

Binding materials other than silicone rubber have been used such as poly(vinyl chloride) (PVC) and thermoplastics such as polyethylene which is particularly suitable since it can be moulded by heating under pressure. Another approach is to simply apply the sparingly soluble inorganic salt to the end of a graphite rod. The graphite rod is usually hydrophobized with Teflon® and mounted in a plastic barrel. The sparingly soluble salt is applied to the end of the graphite and rubbed into it with a glass rod. The solid membrane formed in this way produces an ion-selective electrode with identical response characteristics to other solid state electrodes. It is possible to change from one sensor to another by trimming off the end of the graphite surface with a blade and applying a different inorganic salt.

3 Polymer membrane potentiometric chemical sensors

3.1 The liquid membrane ion-selective electrode

In 1967, a liquid membrane ion-selective electrode was produced which, for the first time, provided the means for the direct determination of the activity of calcium ions in solution. This was of great interest in the biological and clinical sciences because of the importance of calcium in biological fluids. The liquid membrane in this electrode consisted of the calcium salt of di-*n*-decyl phosphate dissolved in di-*n*-octylphenyl phosphonate.

A commercial version of this electrode was produced and is illustrated in Fig. 3.1. The organic liquid saturated the organophilic membrane at the tip of the electrode barrel and acted as the ion-selective membrane which exchanged with calcium ions in the sample solution. It was also shown that changing the organic solvent for the calcium organophosphate salt resulted in different selectivities for ions. For example, the use of a solvent like *n*-decanol in place of di-*n*-octylphenyl phosphonate gave an equal selectivity for calcium and magnesium which resulted in a liquid membrane electrode suitable for the determination of water hardness. This was a great advance in potentiometric sensing and further liquid membrane electrodes followed.

The main requirements for the organic liquid membrane were that it should be immiscible with water, have a low volatility, interact reversibly with the ion of interest and exhibit some degree of charge conduction which, presumably, occurs by the transport of ions by reagent molecules. However, commercial liquid membrane electrodes were generally bulky and rather messy to assemble.

Perhaps the most significant advance in liquid membrane electrodes, other than the original discovery, occurred in 1970 when it was shown that the organic liquid of the liquid membrane ion-selective electrode could be immobilized into poly(vinyl chloride) to produce a polymer film with sensing properties for calcium as good as, if not better than, the liquid membrane itself. This is possible because the reagents and organic liquids used for preparing the liquid membrane are, in general, excellent plasticizers for PVC. Such plasticizers lower the glass transition temperature of PVC and produce homogeneous and flexible films with good mechanical stability. A general 'rule' of thumb is that PVC-based polymer membranes for potentiometric sensors should contain about 70% by weight plasticizer and 30% PVC. The amount of ionophore needed is only about 1% and is included in the amount of plasticizer.

The liquid membrane Ca^{2+} ISE first reported in 1967 (Ross, J. (1967). *Science*, **156**, 1378).

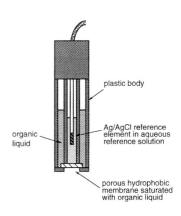

The use of PVC to make sensor membranes originated from the laboratory of Professor J. D. R. Thomas (Moody, G. J., Oke, R. B., and Thomas, J. D. R. (1970). *Analyst*, **95**, 910).

Fig. 3.1 The liquid membrane ion-selective electrode.

plastic body

Ag/AgCl reference element in aqueous reference solution

organic liquid

porous hydrophobic membrane saturated with organic liquid

PVC membranes are simple to make. The components are dissolved in tetrahydrofuran (THF) and the clear solution is poured onto a glass surface. A transparent, flexible membrane is obtained on evaporation of the THF.

The development of PVC membranes has made the original liquid membrane electrode configuration virtually redundant and has led to potentiometric sensors based on polymer membranes being available commercially to sense ions such as calcium, magnesium, water hardness, sodium, potassium, lithium, pH, barium, nitrate, bicarbonate, and ammonium. It should be noted that many of the biologically important ions are included in this list and so potentiometric sensors are used extensively in the analysis of biological fluids. For example, virtually all determinations of potassium in blood serum are carried out using the polymer membrane ion-selective electrode.

PVC-based sensor-membranes all contain a reagent dissolved in a suitable solvent which selectively binds with the ion of interest. Such a reagent is commonly referred to as an ionophore. Much research has been carried out on these types of membranes and there are numerous reports of sensors for many other ions and species. In many ways polymer membranes can still be thought of as highly viscous liquid membranes and are sometimes referred to as 'gelled' liquid membranes' or 'entangled' liquid membranes. In any event, the underlying principles of the way they function are essentially the same. These polymer membranes, generally, have a fairly high electrical resistance (in the order of 1 MΩ) but must conduct charge to function (i.e. in order to measure the potential via Ohm's Law) and, presumably, the charge carrier is the ionophore/ion complex moving within liquid channels in the membrane.

3.2 PVC membranes for sensing calcium

There are two common ionophores used in membranes for sensing calcium ions in aqueous solution and these are shown in Fig. 3.2. The first of these is a cation exchanger, bis(bis(4-(1,1,3,3-tetramethylbutyl)phenyl)phosphato) calcium(II) (BBTP), in which the complexing component is the organo-phosphate anion. This reagent is normally dissolved in dioctylphenyl phosphonate. A typical membrane contains 68–69% plasticizer, 30% PVC, and 1–2% of the exchanger. A change in the composition of the plasticizer will alter, significantly, the selectivity and a divalent ion (equally selective for calcium and magnesium) membrane can be obtained with the composition, 48% n-decanol, 18% dioctylphenyl phosphonate, 33% PVC, and 1% exchanger. This sensor membrane is useful for the determination of water hardness.

The second and more recent ionophore is a neutral carrier, $(-)$-(R,R)-N,N'-bis[(11-ethoxycarbonyl)undecyl]-N,N'-4,5-tetramethyl-3,6-dioxaoctandiamide

Ion-exchanger   Neutral carrier

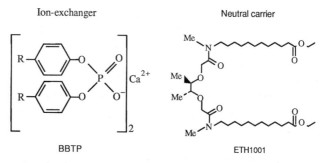

BBTP   ETH1001

Fig. 3.2   Calcium selective ionophores.

(ETH 1001) dissolved in *o*-nitrophenyl-*n*-octyl ether. This reagent is termed a neutral carrier because the ligand is neutral and the complex with calcium is positively charged. A typical membrane composition in this case contains 65.5% plasticizer, 33% PVC, 1% ionophore, and 0.5% potassium tetrakis (*p*-chlorophenyl) borate.

The complexing agent in the case of the ion-exchanger is the phosphoric acid ester, bis(4-(1,1,3,3-tetramethylbutyl)phenyl) phosphoric acid, and it is often assumed that such reagents act as bidentate chelating agents. However, a study of the crystal structure of the calcium complex has shown clearly that calcium ions make only one short contact with the phosphate oxygens and that chelation does not occur. Each calcium ion is complexed with one oxygen of each of four neighbouring phosphates and with two water molecules making the arrangement around each calcium ion octahedral. This arrangement leads to a weaker complex than for the case where the phosphates are chelated and so calcium ion exchange is facilitated. Very strong complexes do not respond in potentiometric sensor membranes.

In the case of the neutral carrier, it is assumed that the calcium ion is complexed with the carbonyl oxygens in the structure and that the long hydrophobic chains wrap around the ion to form what is effectively a cage structure. This complex is positively charged and, to exist in the bulk membrane phase, requires a lipophilic anion to preserve electroneutrality otherwise organophilic anions from the sample will be extracted into the membrane and lead to an anion interference. This is normally overcome by including a lipophilic anion such as the tetrakis(*p*-chlorophenyl)borate ion in the membrane composition.

A typical polymer membrane ion-selective electrode is shown in Fig. 3.3 and the simplicity of construction compared to the original liquid membrane electrode is clearly seen. A membrane can be cast onto a flat surface and a disc cut and glued onto the end of a barrel. A suitable inner reference element is then added. This is the configuration which is normally used in commercial polymer membrane ion-selective electrodes although some manufacturers have modified this for convenience of use. For example, in some cases a sealed module screws onto the end of the barrel. Replacement of the sensor-membrane then simply involves screwing on a new module and so the same barrel can be used with sensing modules for different ions.

As mentioned above, the development of the potentiometric sensor for calcium has, for the first time, given the biologists and the clinical biochemists a method for the direct determination of ionized calcium in biological fluids. Previously, only total calcium could be determined by atomic absorption spectrometry. In certain physiological conditions the amount of ionized calcium in a biological fluid like blood serum may vary widely, whereas the total amount remains reasonably constant. There has also been a considerable amount of work on the development of micro-sized liquid membrane and polymer membrane sensors suitable for measurements in tissues and even in single cells.

The sensor is used in the following electrochemical cell to sense calcium.

$$Hg, Hg_2Cl_2 \mid satd \; KCl \parallel sample \; solution \mid PVC \; membrane \mid$$

$$0.1 \, M \; CaCl_2 \mid AgCl, Ag$$

When the membrane containing the calcium-selective ion-exchanger (organo-phosphate membrane) is placed in contact with an aqueous sample solution a

Many new neutral-carrier sensing reagents originated from the laboratory of the late Professor W. Simon at Eidgenössische Technische Hochschule (ETH) in Zurich.

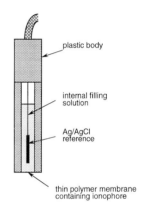

plastic body

internal filling solution

Ag/AgCl reference

thin polymer membrane containing ionophore

Fig. 3.3 A polymer membrane ion-selective electrode.

The inner reference element in the Ca^{2+} ISE would be an Ag metal wire or rod coated with AgCl and immersed in a 0.1 M solution of $CaCl_2$.

The normal range of total calcium in blood serum in adults is 2.20–2.55 mM. Approximately half is present as ionized Ca^{2+}.

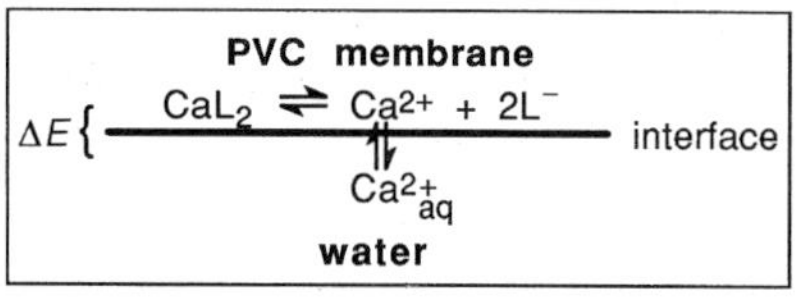

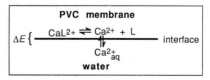

Fig. 3.4   The establishment of the interfacial potential for the calcium/organophosphate membrane.

charge separation process occurs across the membrane/solution interface giving rise to an interfacial potential as illustrated in Fig. 3.4. It is the equilibrium reaction between the ion of interest and the ionophore which controls the interfacial potential. For the calcium/organophosphate complex, calcium ions distribute across the interface, the process being facilitated by the hydration energy of the calcium ion. This leaves negatively charged ligand groups on the membrane side of the interface and the charge separation gives rise to the interfacial potential. The number of calcium ions in the sample solution originally determines the position of the equilibrium and hence the size of the interfacial potential. The cell potential is given by the Nernst equation,

$$E = \text{constant} + 29.6 \log a_{Ca_{aq}^{2+}} \tag{3.1}$$

For the calcium selective membrane based on the neutral carrier, the charge separation process is illustrated in Fig. 3.5. In this case, it is assumed that positively charged calcium/neutral carrier complex ions occupy the interface region of the membrane with negatively charged ions from the solution occupying the surface region on the aqueous side of the interface. This establishes the interfacial potential. The distribution of calcium ions across the interface leaves uncharged neutral carrier on the membrane side of the interface and leads to a change in the interfacial potential. Again, the interfacial potential is controlled by the position of the surface equilibrium and by the number of calcium ions in the sample solution and the cell potential is given by eqn 3.1.

It has become evident from the cases discussed that the response of potentiometric chemical sensors is due in the main to interfacial equilibria. These surface reactions are fast and the change in the interfacial potential is rapid when the activity of the ion of interest in the sample is changed. In many cases, however, a bulk extraction of species into the polymer membrane occurs. Such processes are slower than the interfacial processes and lead to a change in the membrane composition. Consequently, establishment of equilibrium may be slow and drift in the cell potential may be observed. Also, any other ions which can affect the position of the interfacial equilibrium (i.e. complex or exchange with the ionophore) will contribute to the interface potential and cause an interference.

Fig. 3.5   The establishment of the interfacial potential for the calcium/neutral carrier membrane.

As already mentioned, in the case of membranes containing a neutral carrier, it is possible to bulk extract lipophilic anions from the sample solution and this can lead to an anion interference. This effect is counteracted by including in the membrane composition a salt containing a lipophilic anion which will balance the charge on any bulk extracted neutral carrier/cation complex. Such a salt is termed an anion suppressing reagent (ASR). In some cases the effect is very small, e.g. in the PVC/valinomycin membrane for potassium, and the anion suppressing reagent is not needed. In these cases it is found that the bulk extraction of the neutral carrier/cation complex is quite low and the charge on the small amount of extracted complex is satisfied by a small number of residual negative sites left from the manufacture of the polymer matrix or from impurities.

3.3 PVC membrane for sensing potassium

The potentiometric sensor for the potassium ion is another example of the very successful application of a polymer membrane containing a selective ionophore to the determination of an ion of great clinical importance. The potassium ion sensor has replaced the flame photometer for the determination of potassium in clinical samples. In fact, the two most important electrolyte ions in biological fluids, sodium and potassium, are now commonly determined using potentiometric chemical sensors (the sensor for sodium uses a sodium ion selective glass membrane) and commercial instruments are available for this.

The normal range of K^+ in blood serum in adults is 3.5–5.0 mM. The Na^+ concentration is 135–145 mM. Thus, the valinomycin-based sensor for K^+ needs to be very selective for K^+ over Na^+.

The polymer membrane selective for the potassium ion uses the cyclic depsipeptide antibiotic, valinomycin, as the ionophore. The structure of the valinomycin/K^+ complex is shown in Fig. 3.6. This structure contains the sequence of organic acids, D-valine, D-α-hydroxyvaleric acid, L-valine, and L-lactic acid. There is a high degree of intramolecular hydrogen bonding in the molecule between amide NH and CO groups and there are six ester carbonyl oxygen atoms which are directed towards the centre. The potassium ion fits nicely in the cavity created within the interior of the valinomycin molecule and, presumably, is held there through interactions with the six ester carbonyl oxygen atoms. This is the reason for the selectivity of the ionophore for

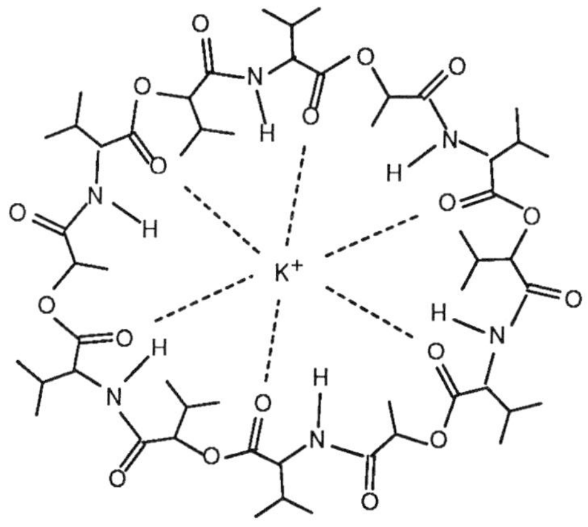

Fig. 3.6 The structure of the valinomycin/K^+ complex.

potassium and the complexation reaction between valinomycin and potassium allows fast and reversible exchange of potassium ions with the sample solution.

Valinomycin is a neutral carrier and so the complex with potassium is cationic. The complex is encapsulated by lipophilic groups which effectively shield the potassium ion from the low polarity membrane medium. However, bulk extraction of potassium into the membrane would require the presence or co-extraction of a lipophilic anion to preserve electroneutrality. As mentioned previously, the extent of the bulk extraction of potassium into the membrane appears to be low and so the membrane composition does not require the addition of an anion suppressing reagent as is the case for the neutral carrier/calcium membrane.

A typical membrane for potassium uses a plasticizer such as di(2-ethylhexyl)sebacate and this membrane is about 5×10^3 times more selective for potassium than for sodium. The membrane composition is approximately 69% plasticizer, 30% PVC, and 1% ionophore. The high selectivity for potassium is demonstrated by the fact that the sensor has been used for the determination of potassium in seawater where the concentration of sodium is 50 times that of potassium.

Commercial versions of the potassium ion-selective electrode are similar to the calcium electrode shown in Fig. 3.3 and the electrochemical cell is written as follows.

Hg, Hg$_2$Cl$_2$ | satd KCl ‖ 10% NH$_4$NO$_3$ ‖ sample solution | membrane |

0.1 M KCl | AgCl, Ag

It should be noted that, in this cell, an ammonium nitrate bridge is inserted between the calomel electrode and the sample. This is to prevent contamination of the sample with potassium ions from the saturated KCl. This is normally done by using what is called a 'double junction' reference electrode with ammonium nitrate solution (10% m/v) in the outer compartment.

The membrane potential for the potassium ion sensor is established by a mechanism similar to that for the neutral carrier/calcium membrane as illustrated in Fig. 3.7. The potential of the electrochemical cell is given by the Nernst equation,

$$E = \text{constant} + 59.1 \log a_{K^+_{aq}} \tag{3.2}$$

3.4 PVC membrane for sensing the nitrate ion

Polymer membrane sensors are also known for the determination of some common anions. In these membranes, use is made of an anion-exchanger as the

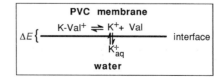

Fig. 3.7 The establishment of the interfacial potential for the potassium/valinomycin membrane.

ionophore and selectivity is usually achieved according to the so-called 'Hofmeister' ion-exchange series, $ClO_4^- > SCN^- > I^- > NO_3^- > Br^- > Cl^- > HCO_3^- > CH_3COO^- > SO_4^{2-} = HPO_4^{2-}$. This is the order of preference of an anion exchange resin for anions. As can be seen, the nitrate ion is reasonably high in this order of selectivity and so a sensor membrane incorporating an anion-exchanger will be successful in the presence of most other ions except those higher in the series. The nitrate ion is of considerable interest since there is a need to be able to determine the amount of nitrate in natural waters, in soils, and in hydroponic nutrient solutions. Much of the nitrate in waters and soils arises from the use of fertilizers. There is also interest in being able to monitor water supplies in remote areas and this has been done using a nitrate ion sensor and a data logging system with a satellite relay to a central laboratory.

There are two types of ionophores which have been successfully used in polymer membrane sensors for nitrate. The first is a quaternary ammonium salt such as tri(n-dodecyl)methylammonium nitrate ($R_3CH_3N^+NO_3^-$, R = $C_{12}H_{25}$) or tetra(n-dodecyl)ammonium nitrate ($R_4N^+NO_3^-$). The plasticizer used with this type of reagent is dibutyl phthalate and a typical membrane composition is 66% plasticizer, 33% PVC, and 1% ionophore.

The second makes use of the nitrate salt of a nonlabile transition metal complex cation such as tris(4,7-diphenyl-1,10-phenanthroline)nickel(II), which has the structure shown in Fig. 3.8. The original liquid membrane nitrate-selective electrode used this reagent dissolved in *p*-nitrocymene.

The following is the electrochemical cell for the nitrate sensor.

Hg, Hg_2Cl_2 | satd KCl || 0.1 M K_2SO_4 || sample solution | membrane | 0.1 M KCl | AgCl, Ag

The double junction reference electrode is used to prevent interference from the saturated KCl solution because the chloride ion, in high concentrations, interferes with the nitrate sensor. Using 0.1 M K_2SO_4 in the outer junction prevents this since the sulfate ion shows negligible interference with the sensor.

The mechanism for the establishment of the membrane potential is shown in Fig. 3.9 and the cell potential is given by,

$$E = \text{constant} - 59.1 \log a_{NO_3^-\,aq} \qquad (3.3)$$

3.5 Other PVC-based sensor membranes

There are numerous other PVC-based membranes in the literature for sensing a whole range of ionic species. Table 3.1 lists some of these sensor membranes. It must be remembered, however, that all these sensor membranes function on similar principles to the ones described above. The response mechanism depends on a charge separation process at the solution/membrane

Three important plant nutrients are nitrogen (as nitrate), phosphorus (as phosphate), and potassium. Nitrate and potassium are easily monitored by potentiometric chemical sensors but there is no successful sensor for phosphate. The form of the phosphate ion depends on the pH since it can exist as the $H^2PO_4^-$, HPO_4^{2-}, or PO_4^{3-} ions.

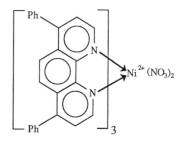

Fig. 3.8 Tris(4,7-diphenyl-1,10-phenanthroline)nickel(II) nitrate.

Many of the reagents and cocktails for making sensors are available from Fluka and are listed in their catalogue *Selectophore* (1996) together with directions for making membranes.

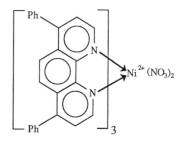

Fig. 3.9 The establishment of the interfacial potential for the nitrate/ion-exchanger membrane.

Table 3.1 PVC-based sensor membranes

Species sensed	Ionophore
Na^+	neutral carrier, ETH 157[a]
pH	tri-*n*-dodecylamine
Li^+	12-crown-4, ETH 1644
NH_4^+	nonactin
Ba^{2+}	Igepal CO-880[b]
Mg^{2+}	neutral carrier, ETH 1117
Zn^{2+}	zinc di-n-octylphenyl phosphate
ClO_4^-	substituted tris (1,10-phenanthroline) ferrous perchlorate
$FeCl_4^-$	aliquat 336 tetrachloroferrate(III)
anionic detergents	quaternary ammonium salt of the detergent anion, e.g. cetyltrimethyl ammonium dodecylbenzene sulfonate
phenobarbital	aliquat 336 salt of phenobarbital
paraquat	paraquat bis (tetraphenyl borate)
promethazine	promethazine tetraphenyl borate
ephedrine	ephedrine tetraphenyl borate
methadone	methadone dinonylnaphthalene sulfonate

[a] Reagents labelled ETH all originate from Eidgenössische Technische Hochschule, Zürich and are available from Fluka.
[b] Nonylphenoxypoly (oxyethylene) ethanol.

interface and, to achieve this, there must be an equilibrium reaction involving the ion of interest and a reagent which can complex or undergo an ion-exchange reaction with the primary ion.

3.6 The coated-wire electrode (CWE)

A conventional polymer membrane ion-selective electrode is shown in Fig. 3.3 and this has an inner aqueous reference system associated with it. The reference element is normally Ag/AgCl and the filling solution contains the ion of interest to establish a constant potential at the inner surface of the membrane. The filling solution also contains the chloride ion to provide a constant potential with the Ag/AgCl. This configuration works well but is not all that easy to miniaturize and to build into devices such as small flow cells. Also, the filling solution must remain in contact with the inner membrane surface and so the electrode must be used in the vertical position. Thus, the report of the 'coated-wire' electrode in 1971 caused considerable interest since it dispensed with the inner reference solution.

The coated-wire electrode originated in the laboratory of the distinguished American analytical chemist, Professor Henry Freiser (Cattrall, R. W. and Freiser, H. (1971). *Analytical Chemistry*, **43**, 1905).

Since this initial report there have been numerous papers describing coated-wire electrodes in many applications and the work laid the foundation for a whole range of so-called 'solid contact' sensors including microelectronic devices such as the ion-selective field effect transistor (ISFET).

In the coated-wire electrode, a polymer membrane of exactly the same composition as described before is cast directly on the surface of a solid substrate. The first coated-wire electrode used a platinum wire as the substrate, but it now known that a less noble metal such as silver or copper is preferable. Other substrates like graphite can also be used as well as other metals such as aluminium.

The original coated-wire electrode as shown in Fig. 3.10 consisted of a piece of platinum wire 1 to 2 cm in length which was soldered to the inner core of a co-axial cable. The end of the wire was beaded by melting gently in a propane/oxygen flame. The bead was then coated with the polymer mixture by dipping into the polymer solution and allowing to dry. The process was repeated to obtain a thicker layer. When dry, exposed metal was bound with paraffin film.

The literature on coated-wire electrodes is now vast and there are many different configurations. All of these, however, have a membrane in direct contact with a solid surface. The electrochemical cell for use with the CWE is the following.

Hg, Hg_2Cl_2 | satd KCl ‖ sample solution | membrane | solid substrate

At first glance, this configuration does not appear to have a charge transport mechanism at the membrane/solid substrate interface. The interface is thus referred to as 'blocked' or 'charge blocked' and the CWE is often described as a 'nonsymmetrical sensor'. This caused considerable controversy in the scientific literature, but the one thing the critics were not able to dispute about CWEs was the fact that numerous workers around the world were able to get them to work. The critics were worried by the fact that in the platinum wire, charge transport is by means of electron flow while in the polymer membrane, ions are transported and there does not seem to be a mechanism to switch from one to the other at the membrane/metal interface.

For any potentiometric sensor to be successful there must be a coupling between the sensor membrane and the rest of the electrochemical circuit which produces a stable potential at the inner surface of the membrane. In conventional ion-selective electrodes this is achieved by bathing the inner membrane surface with a solution containing a constant activity of the ion of interest. The bathing solution also contains an ion which can exchange charge with the reference element. For example, if an Ag/AgCl reference element is used, silver and chloride ions are included in the bathing solution ($AgCl + e^-$ $= Ag + Cl^-$).

The suggestion has been made that in the CWE, a stable potential is achieved at the membrane/metallic substrate interface by a mixed potential mechanism as follows. A redox system exists at the metallic surface which requires both a reducible species and a species which can be oxidized. The reducible species is assumed to be oxygen which diffuses through the membrane and is provided at a constant rate. The species oxidized is the metal in the case of less noble metals such as silver and copper, and impurities in the membrane in the case of platinum. Coupling the oxidation potential with the reduction potential gives a reasonably stable reference potential at the metal/membrane interface.

It is true that CWEs, in general, are inferior with respect to drift and reproducibility than their conventional counterparts with an aqueous reference system built in; however, this is now of little consequence since potentiometric sensors are of more use in flow techniques like flow injection where drift can be easily accounted for in determining the transient signal. CWEs have

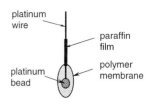

Fig. 3.10 The original coated-wire electrode.

A number of people have tried to introduce a reference layer onto the metal surface prior to depositing the ion-selective membrane.

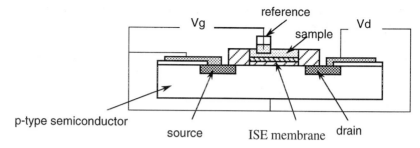

Fig. 3.11 Ion-selective field effect transistor.

considerable advantage in being able to be fabricated into virtually any configuration for inclusion in a flow cell.

3.7   Ion-selective field effect transistor (ISFET)

This device has arisen from the microelectronics field and is an adaptation of the IGFET (insulated gate field effect transistor). Such devices are extremely small (< 1 mm^2) and probably have their most useful application in arrays for multi-species detection. An ion-selective field effect transistor (ISFET) is shown in Fig. 3.11 and basically consists of a p-type semiconductor in which there are two regions of n-type semiconductor which are termed the source (S) and drain (D), respectively. A layer of insulator is deposited on top of this and then an ion-selective membrane, which replaces the metal gate material of an IGFET. The amount of current flowing between the source and drain is dependent on the gate potential.

The sample solution is placed on the ISE membrane and a reference electrode is immersed in the solution. This generates a potential at the membrane surface which is, in effect, the gate potential controlling the flow of current between the source and drain. The amount of current flowing is related to the membrane potential which, in turn, is related to the activity of the ion of interest in the sample solution. It is interesting to note that the insulator layer itself (SiO_2 or Si_3N_4) responds to pH and can be used for this purpose without an additional membrane.

It is common to use ISFETs with circuitry which operates in a feed-back mode (i.e. at a constant drain current) to obtain potentials related by the Nernst equation to the activity of the ion of interest.

It is possible to fabricate ISFETs using most of the membranes which have been described for potentiometric sensing, however, the most studied ISFET and the most successful has been the pH-ISFET. ISFETs for pH measurements use either the insulator layer as mentioned above for their pH response, or a glass membrane which is deposited on the gate. It is also possible to use a polymer membrane which contains a lipophilic amine for pH sensing. ISFETs for pH measurements are commercially available.

3.8   Potentiometric sensors based on gel-immobilized enzymes

These are the first examples of 'biosensors' since they make use of biochemical processes which involve enzyme-catalysed reactions with analytes to produce products which can be sensed by potentiometric chemical sensors. One example is the urea potentiometric sensor shown in Fig. 3.12. This uses a polyacrylamide gel layer containing the enzyme, urease, fixed to the surface of an ammonium ion sensor such as an ammonium ion selective glass electrode. The gel layer is held in place by means of a nylon net. The electrochemical cell used in the measurement of urea in a sample solution is as follows.

Hg, Hg_2Cl_2 | satd KCl || sample solution | urease gel | glass membrane |

0.1 M HCl | AgCl, Ag

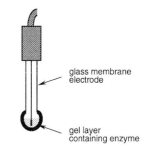

glass membrane electrode

gel layer containing enzyme

Fig. 3.12 The urea potentiometric chemical sensor.

When this sensor is placed in a sample solution containing urea, urea diffuses into the gel and the following reaction takes place.

$$CO(NH_2)_2 + 2H_2O \rightarrow CO_3^{2-} + 2NH_4^+ \qquad (3.4)$$

The ammonium ion formed is sensed by the NH_4^+-sensitive glass electrode and the relationship between the measured potential and the urea concentration is given by the following equation.

$$E = constant + 59.1 \log [\text{urea}] \qquad (3.5)$$

Enzyme-based potentiometric biosensors, in general, tend to be slow to equilibrate since time is required for the analyte to diffuse into the enzyme layer. Some other enzyme-based sensors are listed in Table 3.2.

3.9 Enzyme field effect transistors (ENFETs)

A urea biosensor for diagnostic clinical purposes has also been constructed using an ISFET by chemically binding the enzyme to the gate of the field effect transistor. In this case, the pH is measured instead of the ammonium ion concentration and is dependent on the equilibrium between the carbonate and hydrogen carbonate ions.

$$CO_3^{2-} + H_2O \rightleftharpoons HCO_3^- + OH^- \qquad (3.6)$$

The pH sensitive layer which is deposited on the FET gate initially, is silicon nitride (Si_3N_4). A number of biosensors of this type using ISFETs have been reported and have even been given the name, enzyme field effect transistors (ENFETs). Analytes which have been determined using pH-sensitive ENFETs include urea, glucose, penicillin G, penicillin V, and cephalosporin C.

Table 3.2 Enzyme-based potentiometric biosensors

Analyte	Enzyme	Reaction	Sensor
urea	urease	$CO(NH_2)_2 + 2H_2O = CO_3^{2-} + 2NH_4^+$	NH_4^+
glucose	glucose oxidase	glucose $+ O_2 = H_2O_2 +$ gluconic acid	H^+
L-amino acids	L-AA oxidase	$RCHNH_3^+COO^- + O_2 + H_2O = RCOCOO^- + NH_4^+ + H_2O_2$	NH_4^+
penicillin	penicillinase	penicillin $+ H_2O =$ penicilloic acid	H^+

4 Some practical aspects of the use of potentiometric chemical sensors

4.1 Calibration graphs

As we have seen in the previous chapter, a potentiometric chemical sensor or ion-selective electrode conforms to the Nernst equation (eqn 4.1) when it responds to the activity of an analyte ion (a_i) in an aqueous solution. The constant term takes into consideration all the potentials at the various interfaces in the electrochemical cell which are assumed to remain unchanged when the activity of the analyte changes.

$$E = \text{constant} + RT/nF \ln a_i \qquad (4.1)$$

It is more common in analytical chemistry to express the Nernst factor in millivolts (mV) rather than volts and so the value of 59.1 mV at 298 K is normally used.

A plot of the measured cell potentials for a series of solutions of the analyte against the natural logarithm ($\ln_e$) of the ion activity in each solution will be linear over the range of response of the sensor and the slope of the line will be RT/nF volts (RT/F is often called the Nernst factor, R is the gas constant, 8.314 J K^{-1} mol^{-1}, T is the temperature in Kelvin, and F is the Faraday constant, 96487 C mol^{-1}). It is more usual to use $\log_{10}$ in which case the Nernst factor is $2.303RT/F$ and has a value of 0.0591 V at 298 K (the multiplier 2.303 converts $\ln_e$ to $\log_{10}$). The 'n' in the slope term is the charge on the analyte ion.

Thus, the slope of the Nernst plot will be +59.1 mV for a monovalent cation, +29.6 mV for a divalent cation, −59.1 for a monovalent anion, and −29.6 for a divalent anion. This means that ion-selective electrodes are most sensitive for monovalent ions. They are not particularly sensitive for trivalent ions because the slope is only +19.7 or −19.7 mV/activity decade depending on whether the ion is a cation or an anion.

Some typical plots for the measured cell potential, E, against the logarithm of the activity of the ion of interest, $\log_{10} a_i$, are shown in Fig. 4.1. These are typical of calibration graphs which are obtained when using potentiometric sensors. In practice, it is normal to use solution concentrations instead of activities since concentration is a more meaningful term to the analytical chemist than the activity.

There are several points which should be noted from the response behaviours shown in Fig. 4.1.

1. A potentiometric sensor is said to be behaving in a Nernstian or close to Nernstian fashion if the slope is ±1–2 mV of the theoretical value. Below this range it is sub-Nernstian, above it, it is hyper-Nernstian. Sensors which are outside the Nernstian range can still be useful analytically. Low slopes are more common and sometimes the sensor membrane only needs conditioning

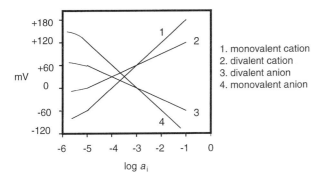

Fig. 4.1 Plots of the cell potential versus the logarithm of the activity for various analyte ions.

in a solution of the ion of interest for an hour or two and then Nernstian behaviour is obtained.

2. The linear activity (or concentration) range is usually between 10^{-5} to 10^{-1} M which is roughly tenths of a mg/L to a few thousand mg/L. This range makes this type of chemical sensor ideal for many environmental problems such as the determination of fluoride in natural waters or the determination of electrolyte ions in biological fluids.

3. There is often curvature in the response below 10^{-5} M. This is due to the sensor becoming insensitive in this region. It can also be due to the effect of an interfering ion on the membrane potential.

4. The potential values (versus an Ag/AgCl or calomel reference electrode), shown on the *y*-axis, are typical of those observed for commercial ion-selective electrodes which have an aqueous inner reference system associated with them. Coated-wire electrodes have potentials which are very positive, i.e. a range of $+50$ to $+300$ mV.

When constructing calibration graphs, the electrodes are first conditioned in a low concentration (0.01 M) solution of the ion of interest for an hour or two and then in de-ionized water for 30 min. Cell potentials are then measured by starting with the most dilute solution first so that these solutions are not contaminated with the more concentrated ones. The electrodes in this case need only be wiped with tissue when changing from one solution to another. If the change is in the other concentration direction, however, the electrodes must be thoroughly rinsed with de-ionized water before placing them in the more dilute solution.

Very gentle stirring of solutions may speed up the equilibration time of the electrodes since this reduces the thickness of the diffusion layer at the surface of the sensor. More vigorous stirring, however, may disrupt the liquid junction associated with the reference electrode and lead to an erroneous value or to drift.

It is essential to use a criterion of stability in order to know when an electrode has reached the 'steady state'. This can be defined in a number of ways but one useful way is to record the cell potential when the value does not change by more than ±0.1 mV over a period of 60 s. Most commercial ion-selective electrodes will reach this criterion in a fairly short time. A more stringent criterion for more accurate work at low concentrations is to require the potential to change by no more than ±0.1 mV over a period of 120 s. With the latter criterion, a commercial fluoride ISE, for example, may take 15 to 30 min to reach the steady state at a fluoride concentration of 0.1 mg/L.

The lower limit of a calibration graph can often be extended by using metal ion buffers for standardization. This is particularly true for electrodes such as the Cu^{2+} ISE which has been used to determine Cu^{2+} in seawater down to 10^{-14} M.

The response time of an ISE is sometimes defined as the time required to reach 95% of the steady state value ($t_{95\%}$).

Some commercial meters for use with ISEs have a built-in facility for determining the stability but usually only apply criteria like the above to do this. This is seen by the operator as a 'READY' prompt. The accuracy required is provided by the operator keying in the appropriate parameter.

4.2   Ionic activities

In order to plot calibration graphs like those in Fig. 4.1, it is necessary to calculate the activity of the ion of interest. This can be done using the following equation.

For a more complete description of the relationship between activity and concentration the reader is referred to Primer 41 in this series entitled *Electrode potentials* by R. G. Compton and G. H. W. Sanders.

$$a_i = \gamma_i[M] \tag{4.2}$$

Where γ_i is the activity coefficient of the ion and $[M]$ is its molar concentration. The ion activity coefficient can be calculated using the following form of the Debye–Hückel equation.

$$-\log \gamma_i = (A\, n_i^2 \sqrt{u})/(1 + Ba\sqrt{u}) \tag{4.3}$$

Where A and B are constants with values of 0.51 and 3.3×10^7, respectively at 298 K. 'a' is the ion size parameter (Table 4.1), 'u' is the ionic strength of the solution, and 'n' is the charge of the ion. This equation holds for dilute aqueous solutions up to an ionic strength of about 0.1 M.

The ionic strength, u, of a solution is calculated using the expression,

$$u = \frac{1}{2}\Sigma(M_{i,j} \cdot n_{i,j}^2) \tag{4.4}$$

Example: Calculate the activity of the sodium and sulfate ions in a 0.01 M solution of sodium sulfate.

(a) The ionic strength of a 0.01 M solution of sodium sulfate is calculated as follows.

$$u = \frac{1}{2}((0.02 \cdot 1^2) + (0.01 \cdot (-2)^2) = 0.03\ \text{M}$$

(b) The activity coefficient of the sodium ion in this solution can be calculated using eqn 4.3.

$$-\log \gamma_i = (0.51) \cdot 1^2 . \sqrt{0.03})/(1 + (3.3 \cdot 10^7 \cdot 4 \cdot 10^{-8} \cdot \sqrt{0.03}))$$
$$\gamma_i = 0.98$$

(c) The activity coefficient of the sulfate ion in this solution can also be calculated.

$$-\log \gamma_i = (0.51) \cdot (-2)^2 . \sqrt{0.03})/(1 + (3.3 \cdot 10^7 \cdot 4 \cdot 10^{-8} \cdot \sqrt{0.03}))$$
$$\gamma_i = 0.91$$

Table 4.1   Ion size parameters (cm)

Sn^{4+}, Ce^{3+}	11×10^{-8}
$H^+, Al^{3+}, Fe^{3+}, Cr^{3+}$	9×10^{-8}
Mg^{2+}	8×10^{-8}
$Li^+, Ca^{2+}, Cu^{2+}, Zn^{2+}, Sn^{2+}, Mn^{2+}, Fe^{2+}, Ni^{2+}, Co^{2+}$	6×10^{-8}
$Sr^{2+}, Ba^{2+}, Cd^{2+}, Hg^{2+}, S^{2-}$, acetate, oxalate	5×10^{-8}
$Na^+, H_2PO_4^-, Pb^{2+}, CO_3^{2-}, SO_4^{2-}, CrO_4{}^{2-}, HPO_4^{2-}, PO_4^{3-}$	4×10^{-8}
$OH^-, F^-, SCN^-, SH^-, ClO_4^-, Cl^-, Br^-, I^-, NO_3^-, K^+, NH_4^+, Ag^+$	3×10^{-8}

4.3 Measuring concentrations

As has been seen, a potentiometric sensor responds to the activity of the ion of interest, however, the analytical chemist is more interested in measuring the concentration of a solution. In order to do this an ionic strength adjuster (ISA) is used. This consists of an electrolyte which is added in the same amount to every solution to adjust the ionic strength to a constant value. This has the effect of making the activity coefficient constant for each concentration of the ion of interest. Thus, eqn 4.2 becomes,

$$a_i = K[M] \tag{4.5}$$

where K is a constant and eqn 4.1 becomes,

$$E = \text{constant}' + RT/nF \ln [M] \tag{4.6}$$

or

$$E = \text{constant}' + 2.303RT/nF \log [M] \tag{4.7}$$

In some cases, for example, when using the calcium ion-selective electrode, the ionic strength of all solutions is adjusted to say 0.1 M simply with an electrolye like NaCl, but in others, the ISA serves other functions as well. A good example of this occurs with the fluoride ion-selective electrode. As seen previously, the only interferent with the fluoride ISE is the hydroxide ion and so the ISA in this case buffers the solutions to pH 5 to 6 to avoid this problem. Also, an indirect interference with the fluoride ISE can occur if metal ions like aluminium(III) or iron(III) are present in the sample. These ions bind the fluoride ion very strongly, and bound fluoride is not sensed by the ISE. In order to measure total fluoride in samples, the ionic strength adjusting solution contains a reagent which complexes the metal ions more strongly than the fluoride ion does and so bound fluoride is released. Reagents which do this are called releasing agents.

The ionic strength adjuster used in the case of fluoride determination is called TISAB (Total Ionic Strength Adjustment Buffer) and there are a number of formulations. One common one contains NaCl to adjust the ionic strength to 0.1 M, sodium acetate to adjust the pH to 5 to 6, and CDTA (*trans*-1,2-diaminocyclohexane-*N,N,N'N'*-tetraacetic acid) to complex the Al(III) and Fe(III).

The response of the F^--ISE becomes very slow after a long period of use in buffer solutions and it has been discovered that the surface of the LaF_3 crystal becomes coated with a layer of $La(OH)F_2$ which is formed by hydrolysis of the crystal surface. This layer can be removed and the fast response restored by polishing the surface using a fine alumina powder.

4.4 Selectivity

Originally, ion-selective electrodes were called specific ion electrodes, but we now know that ISEs are not specific for one ion (although the fluoride ISE is nearly so) but are selective for a particular ion. This means that other ions can interfere by affecting the charge separation process at the membrane/solution interface. For example, with the calcium ISE, other ions, in addition to calcium, are able to complex with the ion-exchanger or neutral carrier in the membrane. In the case of the sulfide ISE, other heavy metal ions form sparingly soluble sulfides and so can influence the silver ion activity at the membrane surface.

Thus, the cell potential measured with a potentiometric sensor, like an ISE, is mainly influenced by the ion of interest (or primary ion) but there will also be a contribution from other ions which can interact with the sensor membrane. Ideally, we would like this contribution to be very small or negligible. In general, a measure of the extent of interference of another ion on the response of the potentiometric sensor is given by the Nicolsky equation (eqn 4.8).

$$E = \text{constant} + RT/nF \ln[a_i + k_{i,j}^{pot} \cdot a_j^{n/x}] \tag{4.8}$$

'n' and 'x' are the charges on the primary and interfering ions of activity 'a_i' and 'a_j', respectively and k_{ij}^{pot} is termed the selectivity coefficient. Further terms are added if there is more than one interfering ion.

Unfortunately, the value for the selectivity coefficient depends on the method used to determine it and so it is difficult to compare values unless all are determined using the same method. However, it is a useful parameter in the qualitative sense, and can be used to gauge the extent of interference another ion might cause on the response of an ion-selective electrode.

A common, rapid, and convenient procedure for the determination of selectivity coefficients is the so-called 'two-point mixed solution method' which is particularly useful for screening a large number of possible interfering ions. The values obtained allow the identification of those ions which are serious interferents.

The procedure consists of measuring the cell potential in a 0.001 M solution of the primary ion alone (E_1) and then in a solution containing 0.001 M of the primary ion plus 0.1 M of the interfering ion (E_2). If the interference is strong then the primary ion concentration used is 0.01 M. In carrying out the calculation of the selectivity coefficient, it is necessary to calculate the activity of both ions in the solutions. The ionic strength will be different in the two solutions.

The selectivity coefficient is then calculated as follows.

$$E_1 = \text{constant} + RT/nF \ln a_i \tag{4.9}$$

$$E_2 = \text{constant} + RT/nF \ln[a_{i'} + k_{i,j}^{pot} \cdot a_j^{n/x}] \tag{4.10}$$

'a_i' is the activity of the primary ion in the 0.001 M solution alone and '$a_{i'}$' is its activity in the mixed solution, 'a_j' is the activity of the interfering ion in the mixed solution. Solving these two equations, simultaneously, gives

$$k_{i,j}^{pot} = \frac{a_i \cdot 10^{(E_2 - E_1)/s} - a_{i'}}{a_j^{n/x}} \tag{4.11}$$

where 's' is the slope of the calibration graph for the primary ion.

In the case of weak interferents, the difference between the two potentials is quite small and some of the reason for this is due to the increase in ionic strength in the mixed solution which lowers the ionic activity coefficient and hence the activity of the primary ion. This decrease in the activity of the primary ion alone produces a decrease in potential which counteracts to some extent the increase in potential due to the interference. For very weak interferents, the potential difference is often too small to measure (i.e. < 0.1 mV). This is associated with k_{ij}^{pot} values less than 10^{-3} and, in these cases, the selectivity coefficient is usually recorded as < 10^{-3}. This is, of course, a deficiency in this approach to the determination of the selectivity coefficient.

Example: The following is an example of how to calculate the selectivity coefficient using the 'two-point mixed solution method'. The cell potential for a calcium-selective ISE (slope $+29.6$ mV/activity decade) in a 0.001 M calcium chloride solution was -20.1 mV. The potential in a mixed solution containing 0.001 M calcium chloride and 0.1 M sodium chloride was -19.8 mV. The task is to use this data to calculate $k^{pot}_{Ca^{2+}, Na^+}$ using eqn 4.11.

The ionic strength of the mixed solution is 0.10 M (eqn 4.4) and it is necessary to calculate the activity coefficients for both Ca^{2+} and Na^+ in this solution. This is done using eqn 4.3 and the values obtained are used to calculate the ionic activities with eqn 4.2. The activity coefficients for Ca^{2+} and Na^+ in the mixed solution are calculated to be 0.40 and 0.77, respectively, which give activities of 4.0×10^{-4} M and 0.077 M, respectively. The value calculated for $k^{pot}_{Ca^{2+}, Na^+}$ is 6.3×10^{-2} which shows that the sodium ion is a rather weak interferent.

A value for the selectivity coefficient which is close to 1 suggests that the sensor is almost equally sensitive to the interfering ion as to the primary ion. A value greater than 1 says that the sensor prefers the interfering ion and that it would strongly interfere with the primary ion response. A problem occurs with multivalent primary ions and monovalent interfering ions since eqn 4.11 produces selectivity coefficients which are unrealistically high because of the power term in the denominator which gives a distorted view of the extent of the interference. However, selectivity coefficients calculated by the 'two-point mixed solution method' are generally only used in a qualitative way and some manufacturers of ISEs even translate them into minimum concentration levels of interferents which can be tolerated in samples.

5 Voltammetric-based chemical sensors

5.1 The technique of voltammetry

The previous chapters have dealt with potentiometric-based chemical sensors involving the measurement of the potential under essentially zero-current conditions which is the basis of the technique of potentiometry. The other important electroanalytical chemistry technique is voltammetry. This involves the measurement of the current flowing in an electrochemical cell as a function of the applied potential.

Many species undergo oxidation or reduction at a potential which is characteristic of the particular species. If the potential is fixed at the value appropriate to the reduction or oxidation of the species being determined, the amount of current which flows is directly related to the concentration of the species. Numerous chemical sensors are used in this mode and are commonly referred to as voltammetric- or amperometric-based sensors.

Many biosensors use a voltammetric transducer and employ highly selective and sensitive biochemical reactions to achieve their selectivity. One amperometric-based sensor which is not a biosensor as such, but which can be incorporated into a biosensor is the oxygen probe.

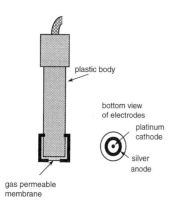

plastic body

bottom view of electrodes

platinum cathode

silver anode

gas permeable membrane

Fig. 5.1 The dissolved oxygen probe.

The solubility of molecular oxygen in water at ambient temperature and pressure is about 8 mg/L.

5.2 The oxygen probe

The oxygen probe (or Clark electrode), which is shown in Fig. 5.1, is an important instrument for the determination of dissolved oxygen particularly in waters associated with environmental studies. The decay of plant and other organic material in lakes and streams can deplete the level of dissolved oxygen to the extent that aquatic life cannot be sustained. The extent to which a waterway has been polluted in this way is determined in terms of its BOD or biological oxygen demand. The BOD is a measure of the rate of dissolved oxygen usage by the organisms in a water sample. The oxygen probe is used to measure the concentration of dissolved oxygen in water.

The term 'probe' is used to describe it because it is a complete electrochemical cell in its own right. The probe has a gold or platinum cathode separated by a plastic casing from a silver anode. A gas permeable membrane covers the electrodes which permits small molecules to diffuse through it. When the probe is immersed in the sample, molecular oxygen diffuses across the membrane into a thin film of electrolyte in contact with the electrodes. The cathode is kept at -800 mV with respect to the silver anode and molecular oxygen is reduced according to the following equation.

The instrument which provides the constant potential is called a potentiostat.

$$O_2 + 2H^+ + 2e^- \rightleftharpoons H_2O_2 \tag{5.1}$$

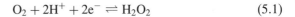

The current flowing in the electrochemical cell is measured and is related to the concentration of dissolved oxygen. Of course, such a probe requires calibration against standards of known concentration.

5.3 The glucose biosensor

The glucose biosensor is perhaps the most successful amperometric-based chemical sensor and has been developed to the stage where a portable unit is available commercially for the determination of glucose in whole blood. This instrument has been designed for personal use by diabetics.

A number of approaches to glucose sensing using amperometry have been reported but most are based on the glucose oxidase (GOD) catalysed oxidation of glucose to gluconic acid as shown in Fig. 5.2.

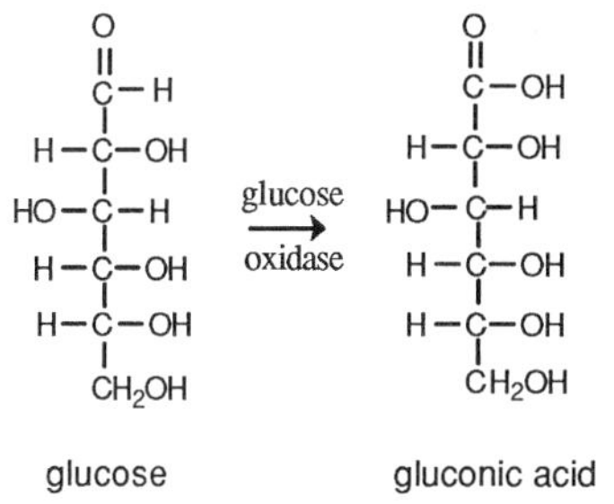

The basic reaction is the oxidation of an aldehyde group to a carboxylic acid group with the production of two electrons.

$$-\overset{\overset{\text{O}}{\|}}{\text{C}}-\text{H}+\text{H}_2\text{O} \rightleftharpoons -\overset{\overset{\text{O}}{\|}}{\text{C}}-\text{O}-\text{H}+2\text{H}^+ +2\text{e}^- \qquad (5.2)$$

The electron acceptor in this reaction is normally molecular oxygen.

$$\text{O}_2 + 2\text{H}^+ + 2\text{e}^- \rightleftharpoons \text{H}_2\text{O}_2 \qquad (5.3)$$

The overall reaction can be written as follows.

$$\text{glucose} + \text{GOD}_{\text{oxidized}} \rightarrow \text{gluconic acid} + \text{GOD}_{\text{reduced}} \qquad (5.4)$$

$$\text{GOD}_{\text{reduced}} + \text{O}_2 + 2\text{H}^+ \rightarrow \text{GOD}_{\text{oxidized}} + \text{H}_2\text{O}_2 \qquad (5.5)$$

There are two ways of making use of this reaction for the quantitative determination of glucose. Firstly, the amount of hydrogen peroxide produced can be determined by anodic oxidation at a fixed potential (e.g. $+600$ mV versus Ag/AgCl) and measuring the amount of current flowing which is related to the amount of glucose in the original sample (the reverse reaction in eqn 5.3). This has been done by chemically attaching the enzyme to the surface of a nylon mesh which is stretched over the surface of a platinum electrode which acts as the anode as shown in Fig. 5.3. The sensor is then placed in a small volume cell along with a Ag/AgCl reference electrode and an auxiliary electrode. Three electrodes are common in voltammetry, the current flows between the platinum and auxiliary electrodes, and the reference electrode monitors the potential. This biosensor system works well for the determination of glucose in the concentration range of 1.8 to 1000 mg/L in foodstuffs such as ice cream, glucose syrup and powder, Horlicks malted milk, molasses, and flour where the interferences are minimal.

A second way to determine glucose makes use of the oxygen probe to determine the decrease in the amount of oxygen in the sample solution after the oxidation of the glucose (eqn 5.5). This can be carried out by spreading a thin layer of glucose oxidase on the surface (or bonding it to the surface) of the gas permeable membrane of the probe. The amount of dissolved oxygen remaining in the sample solution which diffuses across the membrane is determined, and is related to the concentration of glucose.

Fig. 5.2 The oxidation of glucose.

The structure of glucose shown in Fig. 5.2 is the open-chain form, however, in solution glucose exists in two other forms as well called hemiacetals. These are cyclic structures with six-membered rings and they exist in an equilibrium with a small amount of the open-chain structure. The two hemiacetals are labelled α-D-glucose and β-D-glucose. It is the open-chain structure which undergoes the reaction shown in eqn 5.2. As the open-chain form is used up the equilibria shift to produce more.

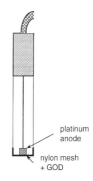

Fig. 5.3 A glucose biosensor.

Three electrodes are used in voltammetry to allow compensation to be made for the resistance of the cell and the electrolyte. This is probably not necessary with voltammetric sensors because the electrodes are very close together and solutions usually contain a high concentration of an electrolyte.

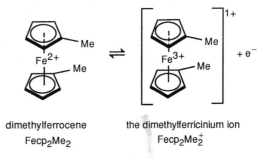

dimethylferrocene

$Fecp_2Me_2$

the dimethylferricinium ion

$Fecp_2Me_2^+$

Fig. 5.4   The oxidation of ferrocene.

5.4   The determination of glucose in whole blood

One important area for glucose determination is in clinical analysis in plasma and whole blood. There is a need for simple, easy to use instruments for people suffering from diabetes to test their sugar levels at home. A number of such instruments use spectrophotometry to determine glucose but recently, a pocket style instrument based on the glucose biosensor has become available commercially for use by patients in the home.

Ferrocene is obtained as an orange solid which sublimes under vacuum and so can be obtained in a very pure form. It is used extensively in electrochemistry because it has a well defined redox potential.

One serious problem with the use of the glucose biosensor in a medium such as blood is the presence of metabolites such as ascorbic acid which can interfere with the determination since they can also be oxidized at the potential of $+600$ mV required to oxidize hydrogen peroxide. Also, the glucose biosensor is sensitive to the oxygen tension (or concentration) of the sample. These problems have been very cleverly overcome by using derivatives of the organometallic compound, ferrocene, to mediate the redox process thus eliminating the need for oxygen in the reaction and the production of hydrogen peroxide. The redox reaction associated with ferrocene and its derivatives is illustrated in Fig. 5.4.

In the glucose oxidase catalysed oxidation of glucose the ferricinium ion acts as the electron acceptor thus eliminating the need for oxygen. The overall reaction is represented in the following equations.

$$\text{glucose} + \text{GOD}_{\text{oxidized}} \rightarrow \text{gluconic acid} + \text{GOD}_{\text{reduced}} \qquad (5.6)$$

$$\text{GOD}_{\text{reduced}} + 2\text{Fecp}_2\text{Me}_2^+ \rightarrow \text{GOD}_{\text{oxidized}} + 2\text{Fecp}_2\text{Me}_2 + 2\text{H}^+ \qquad (5.7)$$

In this process, the potential is held at $+160$ mV which is the potential required to oxidize ferrocene to the ferricinium ion. The oxidation of glucose produces an equivalent amount of the reduced form of GOD which reacts with the ferricinium ion to produce an equivalent amount of ferrocene. This is immediately oxidized back to the ferricinium ion at the set potential and the current flowing is measured. The current flow is proportional to the amount of glucose in the sample. The possible interferences in whole blood are not oxidized at this potential and so do not interfere with the glucose determination.

Many reactions similar to that described above for glucose involving enzymes have been exploited to provide biosensors with an amperometric-based transducer for a range of analytes. The glucose sensor also illustrates some of the interesting chemistry associated with making an enzyme-based biosensor with a reasonably long life-time. This involves chemically binding the enzyme to the sensor surface. In the case of the ferrocene mediated sensor the substrate used is graphite and, firstly, the ferrocene derivative is deposited on the surface from an organic solvent which is allowed to evaporate. Next, glucose oxidase is chemically bonded to the surface using the following reactions. The graphite is heated to 100 °C in air for four hours to activate the surface. This produces carboxylic acid groups on the surface which are then reacted with a 'Woodward' reagent, e.g. 1-cyclohexyl-3-(2-morpholinoethyl)-carbodiimide *p*-methyltoluene-sulfonate, R–N=C=N–R', which binds to the surface through amide linkages. These amide linkages are reacted with glucose oxidase and the enzyme binds through the protein.

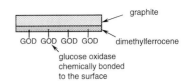

Fig. 5.5 The sensing surface of the glucose biosensor.

Figure 5.5 illustrates the nature of the sensor surface and Fig. 5.6 describes the chemical steps involved in binding the enzyme to the graphite surface. Cases such as this where the surface of an electrode material like graphite (or a metal) has been modified to alter its electrochemical properties are termed 'chemically modified electrodes (CMEs)' and considerable work has been carried out recently in this field.

5.5   The cholesterol biosensor

Another example of the use of an enzyme-catalysed reaction to produce a biosensor occurs with the cholesterol sensor. In this case, the enzyme, cholesterol oxidase, in the presence of molecular oxygen converts cholesterol to cholest-4-en-3-one and H_2O_2. Initially, free cholesterol must be produced from an ester linkage and this is done with the use of cholesterol esterase. Equations 5.8 and 5.9 give the relevant chemical reactions.

$$\text{cholesterol ester} + H_2O \xrightarrow[\text{esterase}]{\text{cholesterol}} \text{cholesterol} + \text{fatty acid} \qquad (5.8)$$

$$\text{cholesterol} + O_2 \xrightarrow[\text{oxidase}]{\text{cholesterol}} \text{cholest-4-en-3-one} + H_2O_2 \qquad (5.9)$$

The sensor operates through the amperometric determination (at a fixed potential) of the H_2O_2 produced using the redox couple hexacyanoferrate(II)/hexacyanoferrate(III). The hexacyanoferrate(II) ion, in this case, acts as the electron donor in the reduction of H_2O_2 according to eqn 5.10. Then the hexacyanoferrate(III) ion formed is reduced at a potential of -50 mV at a platinum electrode and the current flowing is measured and is proportional to the amount of cholesterol in the sample. At this potential there are very few interferents in samples such as blood serum.

$$H_2O_2 + 2[Fe(CN)_6]^{4-} + 2H^+ \rightarrow 2[Fe(CN)_6]^{3-} + 2H_2O \qquad (5.10)$$

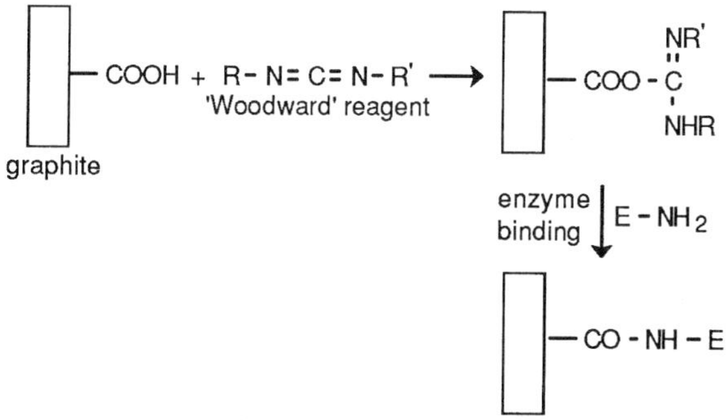

Fig. 5.6   Immobilization of glucose oxidase onto graphite.

5.6   A biosensor to determine the freshness of fish

In addition to their importance in the field of medicine, biosensors are making an impact in the food industry, particularly to monitor food contamination and freshness. One example is a biosensor to monitor the freshness of fish. Obviously, it is crucial in the use and marketing of fish that freshness is ensured. Fish undergoes two stages of degradation, autolysis and putrefaction. During autolysis muscle ATP decomposes to uric acid with the aid of catabolic enzymes. One of the intermediate products formed in this process is hypoxanthine and the concentration of this compound increases with time after the death of the fish. Thus, the determination of the amount of hypoxanthine gives a guide to the freshness of the fish.

Hypoxanthine has been determined using a biosensor consisting of a layer containing the enzyme, xanthine oxidase, in contact with the Clark oxygen probe described in Section 5.2. Xanthine oxidase catalyses the oxidation of hypoxanthine using dissolved oxygen as the electron acceptor. The oxygen probe determines the decrease in the amount of dissolved oxygen in the sample solution and this is related to the amount of hypoxanthine present in the fish sample and, in this respect, the sensor is similar to the glucose biosensor described in Section 5.3.

5.7   Biosensors based on nicotinamide adenine dinucleotide cofactor

The enzyme-based biosensors discussed above all make use of a particular oxidase to facilitate the catalytic oxidation of the analyte. There are, however, around 300 enzymes called dehydrogenases which rely on the presence of a cofactor for their activity and considerable research has gone into investigating these enzymes for use in biosensors. One cofactor is nicotinamide adenine dinucleotide in its oxidized and reduced forms denoted by NAD^+ and NADH, respectively, and for the biosensor to function both the dehydrogenase and the cofactor must be present. A typical dehydrogenase catalysed reaction in the presence of the cofactor is represented by eqn 5.11.

$$\text{substrate} + NAD^+ \xrightarrow{\text{dehydrogenase}} \text{product} + NADH + H^+ \qquad (5.11)$$

The quantitative determination of the substrate requires the measurement of the amount of NAD^+ used up or NADH produced. For a number of chemical reasons it is easier to determine the amount of the reduced form of the cofactor (NADH) formed and this can be done photometrically or electrochemically. The redox relationship between the oxidized and reduced forms of the cofactor is illustrated in Fig. 5.7.

By analogy with the glucose biosensor described above, it would seem feasible to use the oxidation of the NADH formed in the reaction given in eqn 5.11 as a way of determining the amount of the substrate present. This can be done voltammetrically by applying the appropriate potential (which depends on the electrodes used) and measuring the current flowing. Unfortunately, this reaction is irreversible at the high overpotential required and so it is not possible to reuse the biosensor.

There has been considerable research carried out on reagents called modifiers which can be added to the biosensor system to reduce the magnitude

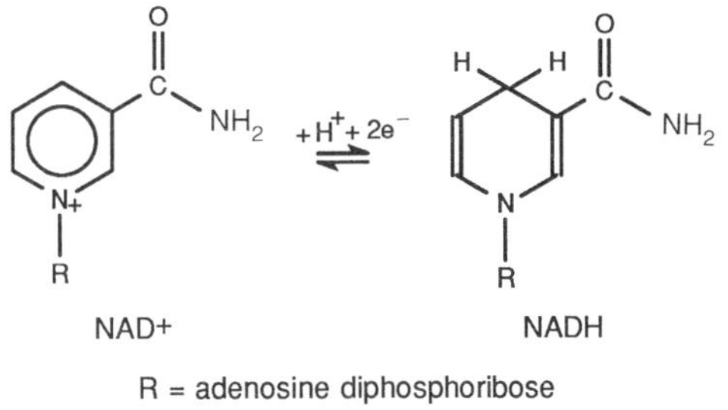

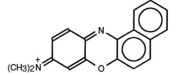

Fig. 5.7 The redox relationship between the oxidized and reduced forms of the cofactor.

of the potential required to oxidize the NADH. These reagents facilitate the transfer of electrons from NADH to an electrode. In this way it is possible to produce reusable voltammetric-based biosensors based on the cofactor in a particular enzyme catalysed reaction.

5.8 A glucose biosensor based on NAD$^+$/NADH

One example of a glucose biosensor based on glucose dehydrogenase uses graphite whose surface has been chemically modified with the 7-dimethyl-amino-1,2-benzophenoxazinium ion (called Medola blue). Compounds such as this bind strongly to graphite and also exchange electrons rapidly with the coenzyme and the cofactor.

The structure of Medola blue is shown in Fig. 5.8 and the glucose dehydrogenase catalysed oxidation of glucose in the presence of NAD$^+$ cofactor is represented in eqn 5.12.

Fig. 5.8 Medola blue.

$$glucose + NAD^+ + H_2O \xrightarrow[\text{dehydrogenase}]{\text{glucose}} gluconic\ acid + NADH \qquad (5.12)$$

The amount of NADH formed is determined at the chemically modified electrode by oxidation using the appropriate potential and measuring the amount of current flowing.

The enzyme and the cofactor need not be immobilized close to the surface of the CME for the system to work, and many flow-based glucose analysis systems are operated in this way. In these, the sample is passed through an on-line reactor containing the enzyme, and NAD$^+$ is added into the sample solution. The solution then passes to the detector containing the CME and the NADH is determined.

For a true biosensor, however, it is necessary to have both the enzyme and the cofactor immobilized on the CME surface. As we have seen previously with the glucose biosensor in Section 5.4, it is not all that difficult to chemically bind an enzyme to a surface like graphite, however, the cofactor is water soluble and so is more difficult to immobilize on a surface. Considerable research is still being carried out on this problem which is being approached in a number of ways. Attempts have been made to bind NAD$^+$ to the enzyme or to couple it with a macromolecule to prevent it escaping to the sample solution. Other approaches use chemically modified carbon paste electrodes with the surface covered with an ion-exchange membrane to protect it. However, the most promising approaches are represented by the following two examples since they are compatible with microelectronic processing techniques, and it is these techniques which will lead to the development of

Microelectronic processing consists of 'thick film' techniques involving screen-printing onto a planar substrate like alumina and fully integrated silicon wafer techniques (called thin film techniques).

cheap, reusable, and reliable biosensors for the challenging monitoring problems of the future as mentioned at the start of this primer.

5.9   A biosensor for determining 'ketone bodies'

In clinical biochemistry the determination of 'ketone bodies' in blood serum is of particular importance and is used as a guide in the treatment of diabetic ketoacidosis. The 'ketone bodies' consist of the compounds, D-3-hydroxybutyrate, acetoacetate, and acetone and are produced by the incomplete fatty acid metabolism in the liver. A biosensor has been developed for monitoring the 'ketone bodies' by determining one of these compounds, D-3-hydroxybutyrate.

The sensor functions by using the D-3-hydroxybutyrate dehydrogenase catalysed oxidation at a CME of D-3-hydroxybutyrate to acetoacetate in the presence of NAD^+ as shown in eqn 5.13.

$$\text{D-3-hydroxybutyrate} + NAD^+ \xrightarrow[\text{dehydrogenase}]{\text{D-3-hydroxybutyrate}} \text{acetoacetate} + NADH \quad (5.13)$$

The sensor, in this case, has been produced using thick film techniques from microelectronics technology. This has been done by incorporating the enzyme, the cofactor, and Medola blue into an aqueous carbon-based paste and screen-printing this onto a substrate, thus producing single use, electrochemical biosensor strips. Obviously, a reference electrode system must also be provided and this can be done using similar screen-printing techniques.

5.10   An ethanol biosensor based on a conducting polymer

A very important advance in producing biosensor membranes is the use of an electrically conducting membrane like polypyrrole which also contains the enzyme, cofactor, and mediator. Such a membrane has been produced for the determination of ethanol by the electrochemical polymerization of a pyrrole solution containing alcohol dehydrogenase, NAD^+, and Medola blue. The membrane can be polymerized onto a metallic substrate to complete the biosensor. The chemical reaction for the determination of ethanol is described in eqn 5.14.

$$\text{ethanol} + NAD^+ \xrightarrow[\text{dehydrogenase}]{\text{alcohol}} \text{acetaldehyde} + NADH \quad (5.14)$$

There is little doubt that techniques like this will be used in the production of intelligent biosensors for many analytes. Such biosensors will be fully reversible and adaptable to many monitoring applications and their production will be compatible with electronics microprocessing techniques.

6 Optical chemical sensors

6.1 Optical measurements

An important category of chemical sensors relies on an optical transducer for signal measurement. Classical ultraviolet–visible (UV/Vis) spectrophotometry is an example of this type of measurement and involves placing a solution of the analyte in the well defined path of an incident beam of light, and measuring the intensity of the transmitted radiation. This is done using a spectrophotometer. The relationship between the intensity of the incident and transmitted radiation is given by the well known Beer–Lambert Law.

$$A = \varepsilon \cdot b \cdot [\mathrm{M}] \qquad (6.1)$$

'A' is the the absorbance $= \log_{10}(P_0/P)$ where 'P_0' and 'P' are the incident and transmitted light intensities, respectively, 'ε' is the molar absorptivity (10^3 mol^{-1}cm^2), 'b' is the pathlength (cm), and [M] is the concentration (mol/L).

This linear relationship applies to monochromatic (a single wavelength) light which is produced by passing the incident radiation through the monochromator of the spectrophotometer. In most instances, some chemistry must be performed on the system such as adding a reagent or reagents which react with the analyte to produce a coloured species which absorbs light at the wavelength of the incident radiation. Conventional ultraviolet–visible spectrophotometry involves a homogeneous solution phase containing the reagents and the analyte.

Optical chemical sensors on the other hand normally involve a two phase system in which the reagents are immobilized in or on a solid substrate such as a membrane which changes colour in the presence of a solution of the analyte. Such an optical chemical sensor has been given the name 'optode'. This term is analogous to the term 'electrode' in electrochemical sensors and initially it was suggested that it should be 'optrode' but, of course, the 'r' only appears in the word 'electron' and so is not appropriate for an optical sensor.

An optode can make use of a homogeneous membrane similar to the membrane described in Chapter 3 for potentiometric chemical sensors or can simply involve a reagent chemically bonded or adsorbed onto the surface of a solid substrate such as glass or plastic. These systems are ideally suited for use in conjunction with optical fibres to yield very versatile sensors for many applications. In cases where an optode membrane is immobilized onto the end (window) of a fibre or a reagent is chemically bound to the window the sensor is referred to as an 'intrinsic' sensor. In other cases, the optical fibre is solely used as a waveguide to transmit light to and from a small sensing surface or a solution and is referred to as an 'extrinsic' sensor. One of the most attractive features of an optode is that it does not require a separate reference sensor as a potentiometric chemical sensor does. Also, it does not suffer from electrical interference (noise pickup) which can be a serious problem with an electrochemical sensor. Optical fibres allow transmission of optical signals over large distances and are therefore ideal for remote sensing. They can also

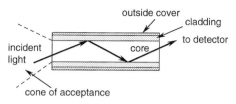

Fig. 6.1   Total internal reflection in a single core optical fibre.

be bent through not too sharp angles and so can be incorporated into a large range of optical sensing devices such as flow cells for continuous monitoring.

6.2   Some properties of optical fibres

It is worth spending a little time describing some of the properties of optical fibres in order to understand how they can be used in chemical sensing. The core of an optical fibre is usually silica, plastic, or glass surrounded by an optical insulator called the cladding which has a lower refractive index than the core. Plastic and glass fibres are much cheaper than silica but are perfectly suitable for sensing in the visible region of the spectrum since neither of these materials absorb radiation in this region. There is also an outside coating for protection, but serving no optical purpose. A section of a single-core optical fibre is shown in Fig. 6.1.

Light enters at one end within the cone of acceptance whose angle is related to the critical angle for reflection at the core/cladding interface. It also leaves at the other end over the same angle. Total internal reflection is the basic mechanism involved in optical fibres and is possible only when the refractive index of the core is greater than that of the cladding and when the incident light enters within the cone of acceptance. Light entering at an angle outside the cone of acceptance will escape from the core. There is a term called the 'Numerical Aperture' which defines the cone of acceptance in terms of the refractive indices of the core and cladding and is a measure of the light gathering capacity of an optical fibre.

Optical fibres can consist of a single-fibre core or bundles of numerous fibres. An optical fibre bundle is said to be 'noncoherent' if the fibres are not aligned within the bundle and can transmit light only. If the fibres within the bundle are aligned (coherent) images can be transmitted as well. A single fibre can transmit light only.

The relationship between the Numerical Aperture and refractive index is $NA = (n_1^2 - n_2^2)^{\frac{1}{2}}/n_0$ where n_1, n_2, and n_0 represent the refractive indices of the core, cladding, and air, respectively $(n_1 › n_2 › n_0)$.

6.3   The evanescent field

In order to reflect light totally within the optical fibre, there must be a zero flux of energy into the optically rarer medium (the cladding). In reality, there is a finite decaying electric field across the interface which extends a small distance into the optically rarer medium. This electric field which penetrates the cladding is called the 'evanescent field'. The depth of penetration of this field depends on the wavelength and, for visible light, is between 100 to 200 nm. This region where the evanescent field exists can be used in optical sensing as described below.

6.4   Optical sensing techniques

Optical sensing techniques that can be used in conjunction with optical fibres are the following.

Absorbance measurements

The simplest form of measurement using optical fibres involves employing the fibres as waveguides to transport light to and from a solution to be analysed. In this case, the analyte is already in a form which can absorb some of the radiation from the incident fibre, or reagents have been added to the solution to convert the analyte to such a form. This application which is an example of 'extrinsic' sensing is the same as conventional ultraviolet–visible spectro-photometry and the analyte solution must be contained in some kind of optical cell (cuvette) or flow cell. The light emanates from a spectrophotometer and is returned to the instrument for interrogation.

The use of optical fibres in conjunction with conventional spectro-photometry has an important application in remote sensing. For example, a remote sensing system has been devised based on this principle which can be used for monitoring methane gas in coal mines. This uses infrared spectrophotometry and the optical cell is located within the mine itself while the infrared spectrometer is outside the mine. In this example, the infrared radiation can be transmitted with optical fibres over a very long distance depending on the depth of the mine.

Remote sensing of this type is also very useful in nuclear processing plants where the radioactivity levels are extremely high. In this application, the strong absorption of light in the visible region of the spectrum by solutions of elements such as uranium and plutonium is used to monitor their levels in processing solutions. Again, optical fibres are used to transmit the light from a spectrophotometer outside the plant to the monitoring site and back again.

In applications where it is necessary to react the analyte with a reagent or reagents to produce the species which interacts with the incident light, it is often convenient to immobilize the reagents in a plastic film or to chemically bind the reagents to the surface of a substrate. The plastic film or chemically bound reagents can be attached directly to the optical window of a fibre. In this case, light transmitted through an optical fibre is passed through the immobilized reagent zone which is also in contact with the analyte solution. The transmitted light is collected by a second optical fibre as shown in Fig. 6.2. The optode membrane changes colour in the presence of the analyte and the intensity of the colour at the wavelength characteristic of the immobilized reagent/analyte interaction is a measure of the concentration of the analyte.

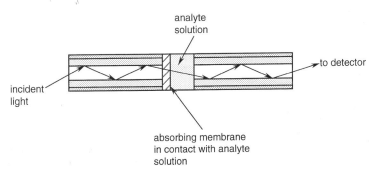

Fig. 6.2　Configuration for absorbance measurements.

Reflectance measurements

Sometimes it is the case that the membrane or medium being used in an optode is opaque and that light is not transmitted through it. Nevertheless, the medium can still interact with an analyte and give rise to a colour change in the medium which is related to the concentration of the analyte. In this case, the light reflected by the opaque medium can be collected and measured. The relationship between the intensity of the reflected light and the concentration is not linear like in absorption measurements, but can be used for the determination of the concentration of an analyte by careful calibration of the system.

One configuration which can be used for diffuse reflectance measurements involves passing light of the appropriate wavelength down an optical fibre and collecting and measuring the light reflected back along the same pathway after interaction with an opaque optode membrane (or medium) attached to the end of the fibre, as shown in Fig. 6.3. In some instances a so-called 'bifurcated' fibre bundle is used in which the fibre bundle is separated into two strands of fibres, one for transmitting light to the sensing area and the other to collect the reflected light.

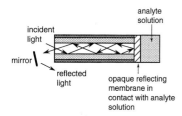

Fig. 6.3 Configuration for reflectance measurements.

Luminescence measurements

When a photon is absorbed by a molecular species an excited state results and radiation is emitted on reverting to the ground state. The term luminescence is applied to the emission of the radiation and includes processes like fluorescence and phosphorescence. Phosphorescence is generally a slower process than fluorescence. Fluorescence spectrometry is an important optical technique in analytical chemistry and is used frequently in chemical sensing. It is a highly sensitive technique and the high sensitivity arises from the fact that the excitation and emission wavelengths are different. This means that the emission signal is measured against an essentially zero background. Fluorescence measurements are of particular interest in biochemical systems where the sensitivity of the optical measurement combined with the sensitivity and selectivity of a biochemical reaction allow extremely low concentrations of an analyte to be determined.

The absorption of light that leads to fluorescence follows the Beer–Lambert Law and so it is not surprising that the intensity of fluorescence is directly related to the amount of absorbed radiation and to the concentration of analyte. The expression relating the intensity of the fluorescence (I_f) radiation to the concentration ([M]), for dilute solutions, is the following.

$$I_f = 2.303 \cdot \phi_f \cdot I_0 \cdot \varepsilon \cdot b \cdot [M] \tag{6.2}$$

'ϕ_f' is the fraction of the total number of photons absorbed that result in fluorescence emission, 'I_0' is the intensity of the excitation radiation, 'ε' is the molar absorptivity, and 'b' the pathlength.

The configuration used for reflectance could also be used for fluorescence measurements where a reagent in the optode membrane reacts with the analyte to produce a product in which fluorescence is excited by incident light at the appropriate wavelength. An alternative configuration makes use of the evanescent field associated with an optical fibre to excite fluorescence. An arrangement for doing this is shown in Fig. 6.4.

In this configuration, the protective coating and cladding are stripped from a small section of an optical fibre exposing the core. A layer or film containing

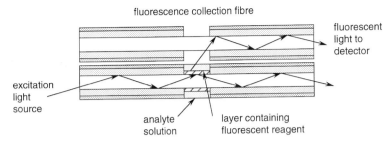

Fig. 6.4 Configuration for fluorescence and evanescent transmission.

a reagent is deposited on the surface of the exposed section of the core. This layer is placed in contact with the analyte which interacts with the reagent in the layer to form a product which fluoresces when excited at the appropriate wavelength. This excitation is achieved using the evanescent field associated with passing light at the appropriate wavelength down the central core of the fibre. The evanescent field penetrates from the core into the active layer and produces fluorescent light. The fluorescent light is collected through the core of a second fibre which has also had its protective coating and cladding stripped as shown in Fig. 6.4.

6.5 Optodes for sensing pH

There has been considerable interest in developing optodes for physiological pH measurements particularly for *in-vivo* monitoring. Optodes present some attractive features for this, namely, the fact that they operate without a separate reference probe and do not have electrical connections. Also, optodes can be easily miniaturized and so are ideal for implanting.

The normal pH range for whole blood is 7.38–7.44.

Optodes for pH measurement generally make use of an acid/base indicator which changes colour about one pH unit either side of the pK_a. Consequently, pH optodes operate over a very narrow pH range (two pH units) compared to the range of say a pH-glass electrode (pH 1–14). This is, however, acceptable for many applications such as pH measurements in biological fluids since the changes in pH are quite small. Obviously, the acid/base indicator to be used in an optode is selected on the basis of its pK_a and the pH range to be measured.

$pK_a = -\log_{10} K_a$ where K_a is the acid dissociation constant. For a monobasic acid (HA), $K_a = [HA]/([H^+] \cdot [A^-])$.

Several pH sensors have been developed based on the use of a hydrogen ion permeable membrane which encapsulates an immobilized colorimetric or fluorescent pH indicator. In one example based on reflectance measurements, phenol red is immobilized on polyacrylamide microspheres (5–10 μm diameter) and packed along with some polystyrene microspheres (1 μm diameter) into a H^+ permeable cellulose acetate dialysis tube attached to the end of an optical fibre as shown in Fig. 6.5. This is immersed in the sample and hydrogen ions diffuse through the dialysis tube.

The polystyrene microspheres are present to scatter the light from the optical fibre to increase the reflectance. Phenol red in its acid form shows a maximum absorption at 560 nm. Light of this wavelength is passed down the fibre, some is absorbed by the indicator with an intensity which depends on the pH and the rest is reflected back to the detector. The difference between the incident and reflected light intensities is related to the pH.

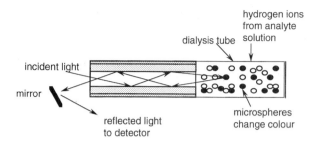

Fig. 6.5 A pH-sensitive optode.

One fluorescent pH indicator which has been used in a similar way to phenol red is HPTS (8-hydroxy-1,3,6-pyrenetrisulfonic acid). HPTS is ideal for physiological pH measurements since its pK_a is 7.3. The acid form of HPTS fluoresces when excited by light of wavelength 405 nm and the intensity of the fluorescent light is related to the pH of the sample.

Another example of a pH sensitive optode makes use of transmission measurements and requires a plastic membrane which is transparent. Membranes which are used in optode sensors are required to extract the analyte from the sample solution into the bulk membrane phase. The extraction equilibrium constant must be high to achieve good sensitivity. This is fundamentally different to membranes used in potentiometric sensors where the potential is established at the solution/membrane interface and bulk extraction of the ion of interest is not a prerequisite for the sensor to function. Of course, bulk extraction of the analyte does take place but the extraction equilibrium constants in this case are usually low and bulk extraction is slow on the electrode response time scale.

pH sensors of this type are normally made using an acid/base indicator whose structure has been modified to make the indicator compatible with the lipophilic membrane. This is usually done by incorporating a long alkyl side-chain into the structure. Such a membrane can be attached to the end of an optical fibre as shown in Fig. 6.2 and the membrane changes colour depending on the solution pH.

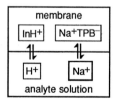

Fig. 6.6 Mechanism for a pH optode—protonated form cationic.

These sensors work by the mechanism shown in Fig. 6.6. Consider an indicator which has been incorporated into a polymer membrane based on cellulose acetate or poly(vinyl chloride), and which has a positive charge in the protonated form and is neutral in the base form. Let us assume that the acid form of the indicator (InH$^+$) is red and the base form (In) is blue. This membrane is transparent and will absorb light at the particular wavelengths associated with the two forms. Electroneutrality in the membrane must be preserved and so a large lipophilic anion such as the tetraphenylborate ion (TPB$^-$) is included in the membrane composition. The membrane colour is controlled by the equilibrium reactions shown which are, in turn, controlled by the solution pH. In the appropriate pH region, the indicator becomes protonated (the extent of which is determined by the actual pH) and, to preserve electroneutrality, sodium ions which are associated with the tetraphenylborate anion must pass into the analyte solution. The reverse reaction takes place in a region of higher pH than this. The absorbance of this membrane is dependent on the ratio of the activity of the hydrogen ion to that of the sodium ion (a_{H^+}/a_{Na^+}) and so the sodium ion activity must be the same in the sample and standard solutions.

In other cases, the acid/base indicator may be neutral in its protonated form and negatively charged in its base form. To preserve electroneutrality in this system, a large cation, such as a quaternary ammonium ion must be included in the membrane composition. In this case, it is the simple anion (e.g. nitrate) associated with the large cation which is transferred to the aqueous phase when the indicator reverts to its base form as shown in Fig. 6.7. The absorbance of the membrane is dependent on the product of the hydrogen and nitrate ion activities, and when making pH measurements, the activity of the nitrate ion must be held constant.

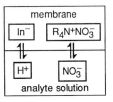

Fig. 6.7 Mechanism for a pH optode—protonated form neutral.

6.6 Optodes for sensing metal ions

The colorimetric determination of metal ions using ultraviolet–visible spectrophotometry has been extensively used in analytical chemistry and there are numerous complexing agents which form coloured complexes with metal ions. Transition metal ions, in particular, form highly coloured complexes due to the *d*-electron transitions which are possible. Other metal ions, however, such as the alkali and alkaline earth metal ions, even though they form complexes with many ligands, do not form coloured complexes. Consequently, ultraviolet–visible spectrophotometry is rarely used for the determination of these ions although some indirect optode procedures are being developed for the colorimetric determination of these ions as described later in this section.

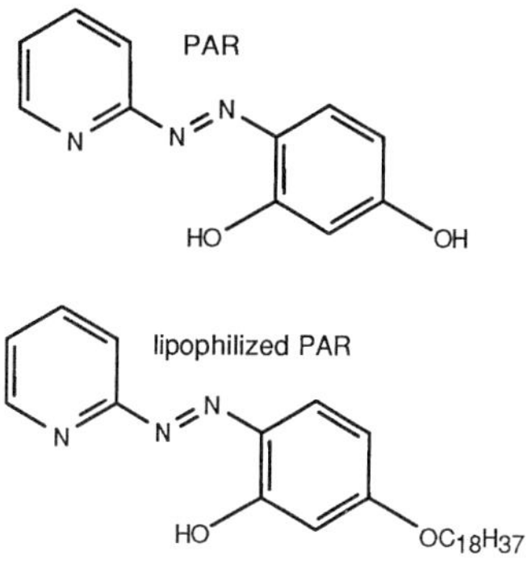

One reagent which is used extensively for the colorimetric determination of transition metal ions is 4-(2-pyridylazo)resorcinol which is commonly called PAR. This reagent has been immobilized into poly(vinyl chloride) to give an optode membrane which has been used to sense zinc ions in solution down to 0.1 mg/L. PAR itself is not sufficiently lipophilic to be compatible with PVC and so it is necessary to introduce a long alkyl side-chain into the structure as described above for pH indicators. The structures of PAR and its lipophilized analogue are shown in Fig. 6.8.

In the presence of zinc ions in aqueous solution, the PVC membrane containing the PAR reagent turns red and the intensity of the colour is related to the amount of zinc. The reaction mechanism is illustrated in Fig. 6.9. Zinc reacts with PAR in the membrane to form a neutral complex (ZnL_2) with the release of two hydrogen ions from the ligand which are transported to the aqueous phase. The absorbance of this membrane is related to the ratio of the zinc ion activity in the sample to the square of the hydrogen ion activity. Consequently, sample solutions need to contain a pH buffer (acetate at pH 4.8) so that the pH does not change during the determination of zinc. This optode

Fig. 6.8 PAR and its lipophilized analogue.

There is a similar reagent to PAR termed PAN, 1-(2-pyridylazo)-2-naphthol, which is also used in optode membranes for sensing metal ions.

PAR forms coloured complexes with many metals and so interferences are a problem with this type of optode. Some discrimination between metals can be achieved by careful selection of the pH and the wavelength of the light source. Metals which form strong complexes with PAR will react at lower pH values than metals which form weak complexes. Also, the wavelength at which maximum absorption of light occurs will depend on the particular metal complex.

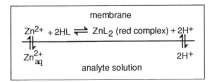

Fig. 6.9 Reaction of zinc with a PAR (HL)-based optode membrane.

membrane takes about 300s to reach 95% of the final signal which demonstrates the relatively slow response compared to electrochemical sensors.

There are other complexing agents which form fluorescent complexes with certain metals. One such reagent is 3,5,7,2',4'-pentahydroxyflavone which is commonly known as Morin. Optode membranes have been prepared by immobilizing this reagent by covalently binding it onto cellulose and measuring the fluorescence intensity when the reagent complexes with metals such as aluminium and beryllium. As mentioned previously, fluorescence measurements provide high sensitivity and an optode made using a Morin-based fluorescent membrane can detect these elements down to 27 μg/L for aluminium and 9 μg/L for beryllium.

As already mentioned, some indirect techniques have been used to develop optode membranes which sense ions of the alkali and alkaline earth metals. The principle of operation of this type of optode membrane is best illustrated by reference to one which has been used to sense Ca^{2+}. In this case, the neutral carrier ionophore which is used in the potentiometric sensor for calcium (see Fig. 3.2) is employed as the selective reagent to bind calcium ions but, of course, the positively charged complex formed in the membrane is colourless. The colorimetric reaction is provided by the lipophilized derivative of the acid/base indicator molecule Nile Blue which is also included in the PVC-based membrane composition. The structure of lipophilized Nile Blue is shown in Fig. 6.10 and this reagent is intensely blue in its protonated form (presumably by attachment of a hydrogen ion to one of the nitrogen atoms in the molecule). The positive charge on the protonated Nile Blue molecule is satisfied by the inclusion in the membrane composition of a large lipophilic anion, in this case, the tetrakis-(3,5-bis(trifluoromethyl)phenyl)borate ion.

This optode membrane which, initially, is intensely blue senses calcium in an aqueous solution by extracting calcium ions into the membrane to form the positively charged complex with the ionophore. To preserve electroneutrality in the membrane, the protonated Nile Blue sheds its proton to the aqueous phase and the intensity of the blue colour decreases. The decrease in intensity of the membrane is related to the concentration of calcium in the aqueous phase. This mechanism is illustrated in Fig. 6.11.

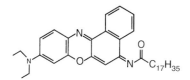

Fig. 6.10   Lipophilized Nile Blue.

6.7   Optode-based biosensor for glucose

An optode has been developed for glucose for use '*in vivo*' (Fig. 6.12) and is based on the following chemistry. It consists of an optical fibre which has a dialysis membrane fitted to one end in a similar way to the pH optode shown in Fig. 6.5. A reagent, concanavalin A, which binds sugars such as glucose and dextran, is chemically bound to the inner surface of the dialysis tube. Initially,

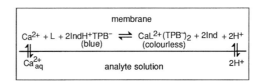

Fig. 6.11   Mechanism for sensing calcium. (L = ionophore; Ind = Nile Blue; TPB = lipophilic anion.)

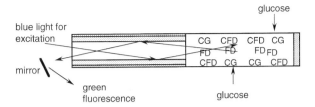

Fig. 6.12 Optode for sensing glucose. (CG = concanavilin A/glucose; FD = fluorescein/dextran; CFD = concanavilin A/fluorescein/dextran.)

dextran labelled with the fluorescent dye, fluorescein, is bound to the concanavilin A. When the dialysis tube is placed in a solution containing glucose, glucose diffuses through the membrane and binds with the concanavilin A displacing an equivalent amount of fluorescein labelled dextran (FD). The freed fluorescein labelled dextran diffuses into the optical path of the blue light being transmitted by the fibre which excites the characteristic green fluorescence of the fluorescein which is transmitted back along the fibre to a detector. The fluorescence intensity is related to the concentration of glucose in the sample.

6.8 Optode-based biosensor for penicillin

This biosensor uses a polymer membrane covalently attached to the end of an optical fibre and also makes use of the fluorescent dye, fluorescein, to detect penicillin. The membrane contains the enzyme, penicillinase, which catalyses the cleavage of the β-lactam ring in penicillin to give penicilloic acid. Fluorescein is also incorporated in the membrane and its fluorescence intensity happens to be pH dependent. Thus, the amount of penicilloic acid produced and therefore the amount of penicillin can be determined from the intensity of the green fluorescence of fluorescein which is excited by blue incident light.

6.9 Optodes for sensing oxygen

It has already been noted in Chapter 5 that the determination of dissolved oxygen in water is important, and the electrochemical probe or Clark electrode for oxygen has been described. There has also been considerable research carried out on the development of optical sensors for oxygen for use in samples as diverse as whole blood, fermentation broth, and natural waters such as lake and river waters. These optodes make use of the ability of molecular oxygen to quench (i.e. attenuate) the fluorescence radiation in an immobilized indicator.

The fluorescence quenching process which occurs by collisional deactivation of the fluorescent indicator by oxygen, is extremely rapid and reversible since no chemical reaction occurs. The decrease in fluorescence intensity is directly related to the amount of molecular oxygen in the sample. The equation which relates collisional quenching to the concentration of the molecule which is responsible for quenching is the Stern–Volmer relationship.

$$I_0/I = 1 + K_q R_0 [Q] \qquad (6.3)$$

Where 'I_0' and 'I' are the fluorescent intensities before and after quenching, respectively, 'K_q' is the bimolecular quenching constant, 'R_0' is the lifetime of the fluorescent molecule in the absence of the quenching molecule, and [Q] is the concentration of the quenching molecule.

One example of an oxygen sensor uses the fluorescent indicator, coumarin 102, immobilized onto an ion-exchange matrix support which is placed against the tip of an optical fibre and encapsulated with a porous Teflon membrane. The assembly is held in place using a length of heat shrink plastic sleeving. The optical fibre is bifurcated (which means it is split into two paths) for transmission of both the excitation and fluorescence light. The fluorescence in the immobilized coumarin 102 is activated by light passing down the optical fibre and the intensity of the fluorescent radiation is measured by collecting the radiation passing back up the fibre. When the end of the fibre is exposed to molecular oxygen some of the fluorescent radiation is quenched and the difference is related to the concentration (or partial pressure) of oxygen.

Complexes of the divalent transition metal ruthenium have fluorescent properties and are often used as the oxygen-sensitive dye in optical sensors for oxygen. An optode sensor for the determination of oxygen in liquids uses the complex tris(4,7-diphenyl-1,10-phenanthroline)ruthenium(II) dichloride as the fluorescent reagent and this complex is adsorbed onto an optical fibre and protected with a membrane coating. This sensor is designed for use in environmental chemistry, clinical biochemistry, and process control, and so must be used without pretreatment of samples. In some samples, there are other luminescent molecules which interfere and so the sensor is designed to operate using the evanescent field mode which has already been described and this minimizes such interferences since fluorescence is restricted to the close proximity of the fibre surface.

6.10   Fluorescent chemosensors

This is a relatively new and rapidly expanding sensor field which had its origins in the need to measure the concentrations of intracellular inorganic ions in cell biology. The term 'chemosensor' has been coined to describe the field, however, it would be more correct to reserve this term for use in describing the actual fluorescent species recognition molecule which binds with an analyte ion, and refer to the overall integrated sensor which incorporates the molecule as an 'optode'. Thus, the few molecules already mentioned above which form fluorescent complexes with metals and H^+ could also be termed 'chemosensors'.

A considerable amount of creative organic chemistry has been carried out to synthesize molecules which bind selectively to an ion and which also possess a fluorescent chromophore whose fluorescence when excited by light (usually from a laser) at the appropriate wavelength is influenced by the bound ion. In this way, the change in fluorescence intensity of the molecule on binding with an ion can be used to determine the amount of ion present.

Fluorescence spectrophotometry is extremely sensitive and so the low concentrations of ions present in intracellular fluid are accessible with this technique. Ions for which chemosensors have been investigated include H^+, Na^+, K^+, Cl^-, Mg^{2+}, Ca^{2+}, and cyclic adenosine-3′,5′-monophosphate. The ion which has been studied more than any other is Ca^{2+} because of its importance in muscle movement in the body. The concentration of extracellular Ca^{2+} is in the range of 1–10 mM, however, intracellular Ca^{2+}

Ca^{2+} is one of the most important inorganic ions in biological systems. As well as controlling muscle movement, it is important in the release of neurotransmitters, in many glands in the endocrine and exocrine systems, as well as in the cells of the immune system. Ca^{2+} also plays an important part at conception and in the development of the embryo. It is not surprising then, that biologists have invested large amounts of effort in finding a Ca^{2+} sensor which can probe the contents of living cells with minimal disturbance to the chemical equilibria which exist there.

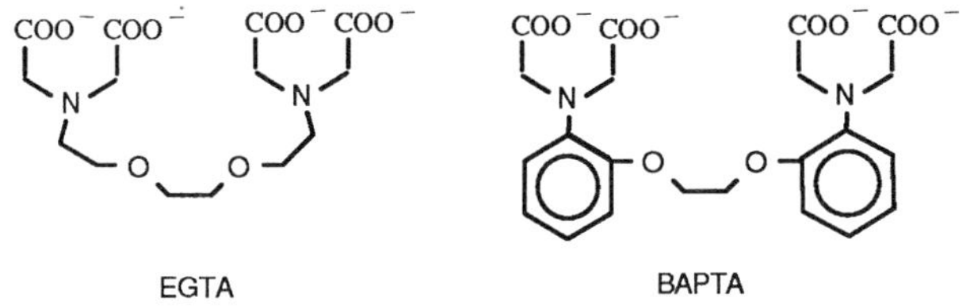

Fig. 6.13   The structures of EGTA and BAPTA.

is much lower than this in the region of 100 nM. The potentiometric chemical sensor for Ca^{2+} easily accesses the higher concentration range but not the intracellular concentration range. Also, the use of a potentiometric chemical sensor is invasive in the sense that the sensor must penetrate the cell wall and this risks introducing higher Ca^{2+} concentrations into the intracellular region.

One problem associated with the determination of Ca^{2+} in intracellular fluid is the relatively high concentration of Mg^{2+} present which is four times the concentration of Ca^{2+}. Thus, a binding agent for Ca^{2+} is used which has a high discrimination against Mg^{2+}. One such reagent which is well known in analytical chemistry is ethyleneglycolbis(β-aminoethylether)-N,N,N',N'-tetra-acetic acid (EGTA) which is used to titrate Ca^{2+} in the presence of Mg^{2+}. This reagent does not have fluorescence properties itself but its structure can be modified to introduce a fluorescent chromophore. One compound which has been used extensively in intracellular measurements of Ca^{2+} is 1,2-bis (o-aminophenoxy)ethane-N,N,N',N'-tetraaceticacid (BAPTA). This reagent retains the selectivity for Ca^{2+} but introduces the aromaticity required to confer fluorescence properties on the molecule which are modulated by binding to Ca^{2+}. The structure of BAPTA is shown in Fig. 6.13 and can be compared with that for EGTA which is also shown.

Other conjugates of BAPTA have been synthesized which have much higher fluorescence quantum yields and these are based on the incorporation of the aromatic ring in BAPTA into conjugated heterocyclic systems. The introduction of the fluorescent dye into the intracellular fluid requires, firstly, masking the carboxylic acid groups as the acetoxymethyl esters and, in this form, the reagents passively cross the plasma membrane. The ester groups are cleaved within the cell and the reagent binds with Ca^{2+}. The fluorescence emanating from the cell is probed using especially developed imaging technology and allows real-time changes in Ca^{2+} levels within the cell to be determined.

Considerable effort is being applied by organic chemists to synthesize fluorescent molecules with selective binding properties for a range of analytes of interest in fields such as cell biology, medicine, environmental analytical chemistry, and industrial chemistry. These reagents can be immobilized on optical fibres either in membranes or by directly binding them to the fibre surface and a number of useful optodes have arisen from this field of research.

One such optode for sensing Ca^{2+} has been constructed by immobilizing a derivative of BAPTA in the hydrophilic polymer, acrylamide, and attaching it to the end of an optical fibre. Fluorescence is excited by an argon laser lasing at 488 nm by passing the light down the fibre. The sensor is dipped into a sample solution and the intensity of the fluorescent light emanating from the membrane is measured through the objective lens of a microscope as indicated

EGTA is an analogue of ethylenediaminetetraacetic acid (EDTA). These are hexadentate chelating agents and bind to Ca^{2+} to form an octahedral complex anion of charge -2. The formation constant of the Ca^{2+} complex in each case is considerably higher than for the Mg^{2+} complex, which is the reason for the selectivity for Ca^{2+}.

in Fig. 6.14. An optode which uses a similar measuring technique has been used to determine glucose but, in this case, employs the quenching of the fluorescence of tris(1,10-phenanthroline)ruthenium(II) chloride by molecular oxygen to determine the amount of oxygen used in the glucose oxidase catalysed oxidation of glucose which is then related to the concentration of glucose present in the sample.

Another optode which has been suggested is intended for use in the continuous monitoring of K^+ in the extracorporeal blood of patients undergoing open-heart surgery. This system uses 4-methylcoumaro-[222] cryptand (MCC222) as the fluorescent reagent and this binds K^+ selectively. The fluorescence intensity is thus related to the concentration of K^+ in the sample. The structure of MCC222 is shown in Fig. 6.15. Cryptands and crown ethers are known to bind with alkali metal ions. The level of K^+ in blood serum is 3.5–5.3 mM and Na^+ is 135–148 mM and so the reagent needs to show considerable discrimination against Na^+. It is important to note that the K^+ concentration range in blood is easily accessible with the valinomycin-based potentiometric sensor and so the potentiometric sensor is a serious competitor with the optode for this particular application. The advantage with the optode, however, is that it does not require a reference electrode.

6.11   Optodes which use the fluorescence properties of NADH

In Chapter 5 we saw how dehydrogenases can be used in biosensors in conjunction with the cofactor nicotinamide adenine dinucleotide (NAD^+/ NADH) by determining, electrochemically, the concentration of the reduced form of the cofactor present. Another property of the reduced form of the cofactor (NADH) is that it fluoresces when excited by light of the appropriate wavelength. Thus, it is possible to determine the amount of NADH present by fluorescence spectrophotometry.

This approach has been used in an optode biosensor for the determination of lactate. In this biosensor, lactate dehydrogenase is immobilized on a nylon membrane attached to the tip of an optical fibre. The enzyme-catalysed reaction leads to the oxidation of lactate to pyruvate in the presence of the cofactor NAD^+ which is added to the sample solution. Lactate in the sample diffuses to the biocatalytic layer where it is oxidized producing NADH at the tip of the fibre. The NADH fluorescence is excited by light of the appropriate wavelength transmitted through the fibre and the resulting fluorescence intensity is measured and related to the amount of lactate in the sample. It is possible to couple a large number of dehydrogenase-catalysed reactions in this way with the fluorescence of the NADH formed.

Another example is the determination of ethanol using alcohol dehydrogenase in the presence of the cofactor. In this case, alcohol in the sample is oxidized to acetaldehyde with the formation of an equivalent amount of NADH.

There is great interest in monitoring in real time the K^+ concentration in blood in patients undergoing open-heart surgery. At present, the K^+ level is usually determined by a batch procedure which involves sending a blood sample to the laboratory which consumes valuable time.

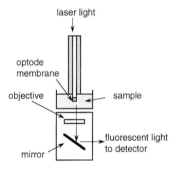

Fig. 6.14   Optode system for sensing Ca^{2+} by fluorescence.

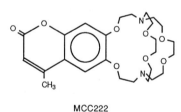

MCC222

Fig. 6.15   4-Methylcoumaro-[222] cryptand.

7 Calorimetric sensors

7.1 Heats of chemical reactions

Chemical reactions produce heat and the quantity of heat produced depends on the amounts of the reactants. Thus, the measurement of the heat of reaction can be related to the amount of a particular reactant. Measurements of the heats of reaction form the basis of the field of calorimetry and this has provided a class of chemical sensors which has gained considerable importance.

Calorimetric sensors, like all other chemical sensors, have a region where a chemical reaction takes place and a transducer which responds to heat. There are three classes of calorimetric sensors which are of particular importance. The first uses a temperature probe such as a thermistor as the transducer to sense the heat involved in a reaction on its surface. The second class of calorimetric sensors are referred to as catalytic sensors and are used for sensing flammable gases. The third class are called thermal conductivity sensors and these sense a change in the thermal conductivity of the atmosphere in the presence of a gas. This is the basis of the thermal conductivity detector which has been used for many years in gas chromatography.

7.2 Thermistor-based sensors

A thermistor is a very sensitive device for measuring changes in temperature and is based on the decrease in electrical resistance (approximately 4–7%/°C) of high-temperature sintered metal oxides (BaO/CaO, transition metal oxide) with temperature. A thermistor is very useful for measuring temperatures with an accuracy of ±0.005 °C. Thermistors can be constructed in many different sizes and shapes but the most convenient one for sensor use consists of a bead covered with a protective coating of glass as shown in Fig. 7.1. The resistance and hence the temperature is generally measured using a Wheatstone bridge circuit. Often a second thermistor is used as a reference.

The high sensitivity to small changes in temperature shown by a thermistor can be employed to sense the small amounts of heat evolved in chemical reactions. This is how thermistors are used in microcalorimetry where a chemical reaction is studied in a bulk solution phase. For sensor application, however, selectivity is required and this is achieved by carrying out a chemical reaction at or near to the surface of the thermistor which, preferably, only involves the analyte of interest.

Enzymatic reactions are highly selective and can involve a significant heat change and there are many reports of the use of thermistors as detectors for such reactions. An obvious advantage of a calorimetric sensor is that it can be used in turbid and highly coloured solutions and there is no requirement to couple the reaction with another reaction to give a colour change as is often required by an optical transducer. Urea and glucose are the most studied analytes and examples of calorimetric sensors for these are discussed below.

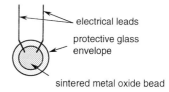

electrical leads

protective glass envelope

sintered metal oxide bead

Fig. 7.1 A bead thermistor.

A Wheatstone bridge is a sensitive instrument for measuring resistance. It consists of two resistance arms, one of which contains two resistances of known value (R_1 and R_2) and the second which contains the unknown resistance and a variable resistance (R_3 and R_4). R_4 is varied until the two arms are in balance in which condition, $R_1/R_2 = R_3/R_4$, thus enabling R_3 to be obtained.

There are generally two approaches to the use of thermistors in calorimetric sensing. One involves incorporating a thermistor in a detector cell to measure the temperature change after the analyte solution has passed through a bed of immobilized enzyme. Although detector systems like this can be adapted to the determination of several analytes they need a large quantity of enzyme. The second approach involves the attachment of the enzyme to the surface of the thermistor itself. The sensor in this case can be miniaturized and easily incorporated into a flow-based analysis system. It is this second type of sensor which is described in the following examples.

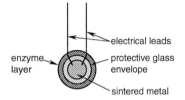

enzyme layer
electrical leads
protective glass envelope
sintered metal

Fig. 7.2 Calorimetric enzyme-coated sensor.

7.3   Calorimetric glucose and urea sensors

In these sensors, the surface of a thermistor is coated with a mixture of albumin and enzyme crosslinked with glutaraldehyde. A second thermistor is coated with only albumin and acts as a reference. When the thermistor coated with enzyme comes in contact with a solution of the analyte (glucose or urea), an exothermic reaction occurs and the thermistor registers the heat change. The amount of heat generated is related to the quantity of analyte in the sample solution. A calorimetric enzyme-coated sensor is illustrated in Fig. 7.2.

The reaction taking place in the glucose case is represented by eqn 7.1,

$$\text{glucose} + O_2 + H_2O \xrightarrow[\Delta H = -80 \text{ kJ/mol}]{\text{glucose oxidase}} \text{gluconic acid} + H_2O_2 \qquad (7.1)$$

and in the urea case by eqn 7.2.

$$CO(NH_2)_2 + H_2O \xrightarrow[\Delta H = -6.6 \text{ kJ/mol}]{\text{urease}} CO_2 + 2NH_3 \qquad (7.2)$$

It can be seen that the glucose sensor is far more sensitive than the urea sensor because of the much larger enthalpy (heat of reaction) involved. Such a sensor has been used for the determination of glucose down to about 2 mM.

7.4   Catalytic gas sensors

An entirely different type of calorimetric sensor to the thermistor-based examples just described has found extensive use in the detection of flammable gases (methane, ethane, propane, butane, carbon monoxide, and hydrogen) and vapours (petrol, organic solvents) in air. The need for the detection of flammable gases is, of course, an old problem and stems from the days when miners used the flame safety lamp to detect methane in coal mines. Nowadays, there are hand held meters which can detect flammable gases in the range of 0–5% in air.

The detection of small amounts of flammable gases and vapours in the atmosphere is extremely important in industry and in the home particularly in regions where natural gas is the main source of energy for cooking and heating. In industry, flammable and volatile solvents are used extensively and the atmosphere needs to be monitored. Small amounts of a flammable gas and volatile solvents mixed with air can form explosive mixtures and it is important to sense the presence of the gas or vapour before it reaches a dangerous level. The term which defines the explosive concentration of a flammable gas is the LEL or 'lower explosive limit' and, for most gases, this lies between 1–5% on a volume basis. Thus, measurements are usually quoted in terms of the % LEL where, for example, 100% LEL for methane is 5% v/v.

A common group of sensors used for detecting flammable gases are the so called catalytic sensors. The principle of operation of a catalytic sensor involves the controlled combustion of the flammable gas in air and the measurement of the quantity of heat evolved in the process. For example, the combustion of methane in air produces 800 kJ/mol of heat which is quite large, and so the measurement of this heat provides a very sensitive way to determine the amount of gas present.

Any sensor needs to respond rapidly, and so to speed up the combustion process a catalyst is used. Thus, a catalytic gas sensor requires a heater to keep the sensor at the temperature to combust the gas, a catalyst to assist in the combustion process, and a device to measure the heat evolved during combustion. The most convenient way to do this is to use a coil of wire as the heater and to also make use of the temperature dependence of the resistance of the wire to measure the heat evolved. Platinum makes an excellent resistance thermometer and it also happens to act as a catalyst in the oxidation of hydrocarbons. Thus, the first catalytic gas sensors used a coil of platinum wire which was heated by passing a current through it to give the temperature for combustion of the gas, which occurred at the surface of the platinum. The heat evolved increased the temperature of the wire and hence its electrical resistance. The change in temperature, of course, is related to the amount of gas combusted. This mode of measurement for using the sensor is referred to as the non-isothermal mode.

The electronics involved in the measurement system sometimes operates in a feed-back mode and, in this case, would involve a decrease in the current required to heat the platinum wire in order to compensate for the temperature rise due to the combustion process. This is called the isothermal mode of operation and the change in current is the parameter which is measured and is related to the temperature change caused by the combustion of the gas and hence the quantity of gas.

Platinum is a poorer catalyst than other metals like palladium and rhodium and needs to operate at a fairly high temperature (1000 °C). This leads to eventual loss of platinum and a reduction in the thickness of the wire. Thus, other forms of the catalytic gas sensor evolved.

7.5 The pellistor

This gas sensor is based on the same principle as before and retains the platinum coil as the heater and temperature sensor (resistance thermometer) but uses a more efficient catalyst for the combustion process. The catalyst used is palladium which is in a finely divided form to increase the surface area and this increases the catalytic efficiency even more. These changes allow the sensor to operate at around 500 °C for hydrocarbons like methane. A diagram of a pellistor is shown in Fig. 7.3. Firstly, the platinum coil is encapsulated in a refractory bead which is about 1 mm in size. Then, finely divided palladium in a thorium oxide matrix is deposited on the surface of the bead.

One serious problem suffered by catalytic gas sensors is their susceptibility to poisoning by other gaseous molecules with associated loss in sensitivity. For example, the sensor can be poisoned by organosulfur and organochlorine

Methane burns in air according to the following exothermic reaction producing 800 kilojoules of heat per mole of methane.

$$CH_4 + 2O_2 \rightarrow CO_2 + 2H_2O$$

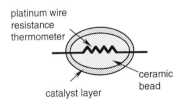

Fig. 7.3 A pellistor gas sensor.

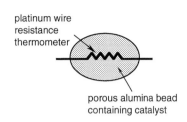

platinum wire
resistance
thermometer

porous alumina bead
containing catalyst

Fig. 7.4   A poison resistant pellistor gas

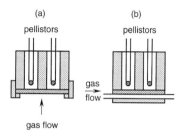

(a)   (b)
pellistors   pellistors

gas
flow

gas flow

Fig. 7.5   Sensing heads used with
pellistor gas sensor.

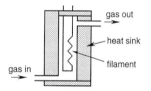

gas out

heat sink

filament

gas in

Fig. 7.6   A thermal conductivity
detector for use in gas chromatography.

compounds, alkyllead additives in petroleum products, and other vapours which are present in industrial plants. Because of this, considerable research has gone into producing pellistors and sensing systems with a lower susceptibility to poisoning. Charcoal-based filters have been used which remove the above poisons, however, such filters will absorb some hydrocarbons as well.

The most satisfactory answer has been to construct pellistors in which the platinum wire is surrounded by a porous alumina bead containing a larger amount of the finely divided catalyst as shown in Fig. 7.4. The available surface area of catalyst is much greater in this case but such pellistors are mechanically weaker than the original types.

The instrumentation associated with gas sensors is reasonably simple and portable, and hand-held instruments are common. In these, the pellistor is incorporated into a small sensing head (Fig. 7.5) usually along with a second but catalytically inert pellistor which is connected to the reference arm of a Wheatstone bridge. A sample of the air being tested diffuses through a metal sinter and comes in contact with the sensor (Fig. 7.5(a)). Alternatively, air can be drawn along a tube into the sensing head using a pump (Fig. 7.5(b)). Gas sensors are fast in response and readings can be obtained in less than 20 s.

7.6   The thermal conductivity sensor

This sensor, unlike the other two already discussed, does not involve a chemical reaction taking place at the sensor surface. Instead, use is made of the thermal conductivity of a gas. This type of sensor has been successfully used for many years as a detector in the analytical chemistry technique of gas chromatography (GC or GLC) and as a gas sensor in industry. The principle of operation involves heating a filament made of tungsten, tungsten/rhenium, or nickel/iron alloys to less than dull red heat (about 250 °C). Heat is lost from the filament to the surroundings and this is dependent on the thermal conductivity of the surrounding gas. The thermal conductivities of gases vary widely and the temperature of the filament will change according to the nature of the gas and its concentration. The change in temperature of the filament is detected as a change in its electrical resistance in the same way as for the other calorimetric sensors. A typical thermal conductivity gas detector as used in gas chromatography is shown in Fig. 7.6.

The thermal conductivity sensor is used in situations where the concentration of gas is likely to be relatively high. Because it does not depend on a chemical reaction for its operation, it can be used in an inert gas environment such as for monitoring the flammable gas content of a container after it has been flushed with nitrogen. It can also be used for detecting inert gases themselves, such as nitrogen, helium, argon, and carbon dioxide. A portable hand held instrument would have a sensing head similar to that shown in Fig. 7.5. It can be seen that the thermal conductivity sensor has special uses of its own which are different to but which complement those of a catalytic gas sensor.

8 Solid-electrolyte and semiconductor gas sensors

8.1 Chemical sensors in the automobile industry

The modern motor car is an excellent example of a mass-produced product which has come under very close scrutiny and control from governments because of concern for our environment and for safety. The car has the potential not only to destroy our environment by pollution of the air we breathe by exhaust gases, but it also consumes enormous natural energy resources in its thirst for petroleum products. A third concern is, of course, safety which has enormous consequences for human survival. The industry has responded to these concerns by designing 'smart cars' equipped with sensors and computers to improve their safety and efficiency.

Both physical and chemical sensors are used in cars. Temperature sensors are used to monitor such things as the engine temperature, the temperature of the brakes, and for cabin climate control. Force sensors determine when to actuate the air bag and when to correct for steering and braking errors. Chemical sensors are used for controlling the engine performance and efficiency so that the emission of noxious gases and fuel consumption are at a minimum. An indication of the extent of the use of sensors in the modern automobile engine is given in Fig. 8.1.

In this chapter, the types of chemical sensors used in automobiles are discussed. Of course, these sensors find uses in many other industries as well and mention is made of these. Two types of sensors are discussed which are classified as 'solid-electrolyte' and 'semiconductor' sensors and arise from the

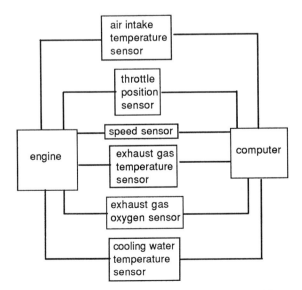

Fig. 8.1 Sensors controlling the engine performance in a modern automobile.

microelectronics and ceramics fields. The largest use of both of these chemical sensor types is in the automobile industry.

8.2   The solid-electrolyte Galvanic cell oxygen sensor

The term Galvanic cell is named after Luigi Galvani an eighteenth century Italian physiologist who pioneered work in the field of electrophysiology. A Galvanic cell (sometimes called a voltaic cell or, in the oxygen sensor case, a concentration cell) is an electrochemical cell in which a spontaneous redox reaction produces electricity.

The most successful example of a solid-electrolyte-based sensor is the oxygen sensor. Like the oxygen probe discussed in Chapter 5, this sensor uses electrochemical principles for its operation, however, these principles are different to those of the oxygen probe. The oxygen probe is an example of a voltammetric-based sensor which is used in aqueous media, whereas the oxygen sensor discussed here is based on a Galvanic cell (oxygen concentration cell) which uses solid-state electrochemistry in its operation.

The EMF of this cell is generated by the difference in the concentration of molecular oxygen at two electrodes under constant temperature and pressure conditions. The solid electrolyte used is a ceramic material which is able to transport the O^{2-} ion. Zirconium dioxide (zirconia) is one such material which has a high oxygen ion (inorganic chemists would call this the oxide ion) conductivity above 500 °C. Some molten metal oxides also transport the oxygen ion but their use poses some difficulties because of their corrosive properties.

The transport of the oxygen ion in zirconia is by means of a lattice defect mechanism by which anions move from one positively charged hole in the structure to another. The oxygen ion conduction can be increased by increasing the number of positively charged holes. This is done by including in the structure another metal oxide like CaO, MgO, or Y_2O_3.

A typical oxygen ion conducting ceramic is made by slowly evaporating an aqueous suspension of $ZrO(NO_3)_2$ and $CaCO_3$ which yields a fine white powder. The nitrate and carbonate ions are decomposed by heating the powder at 700–800 °C to give a mixture of the oxides. The powder is then pressed into the required shape and sintered at 1500 °C yielding a white ceramic material. This final ceramic contains about 8% (m/m) CaO, 89% ZrO_2, plus traces of other oxides of other elements such as Si.

In a mixture of gases, the total pressure is the sum of the partial pressures of the individual components. Thus, the partial pressure of a gas is the pressure it would exert if it were present alone. The term 'partial pressure' is often used instead of concentration to indicate the quantity of a component in a mixture.

The ceramic is built into an oxygen concentration cell in which it acts as the oxygen-ion conductor and also serves as the separator between two compartments with gaseous molecular oxygen having a different partial pressure in each. The Galvanic cell can be written as follows.

$$p_{O_2}^1 | \text{ Pt } | \text{ ZrO}_2/\text{CaO } | \text{ Pt } | p_{O_2}^2 \tag{8.1}$$

where $p_{O_2}^1$ and $p_{O_2}^2$ are the oxygen partial pressures in compartments 1 and 2, and the ceramic is coated with porous platinum electrodes.

The above cell generates an open-circuit potential due to the following half-cell reactions at each platinum electrode, for the case where $p_{O_2}^2 > p_{O_2}^1$.

$$\text{At electrode 2} \qquad 4e^- + O_2(p_{O_2}^2) \rightarrow 2O^{2-} \tag{8.2}$$

$$\text{At electrode 1} \qquad 2O^{2-} \rightarrow O_2(p_{O_2}^1) + 4e^- \tag{8.3}$$

The oxygen ion generated at electrode 2 is transported through the ceramic material to electrode 1 where it is discharged as molecular oxygen. The overall reaction is given by eqn 8.4.

$$O_2(p_{O_2}^2) \rightarrow O_2(p_{O_2}^1) \tag{8.4}$$

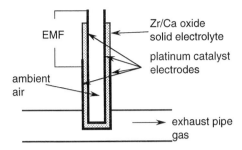

Fig. 8.2 A solid electrolyte oxygen sensor for use in an automobile engine.

The potential of this cell is given by the following form of the Nernst equation.

$$E = RT/4F \ \ln(p_{O_2}^2/p_{O_2}^1) \qquad (8.5)$$

In the operation of the sensor, the reference side of the cell $(p_{O_2}^2)$ is usually atmospheric oxygen and so the measurement of the potential of the cell allows the partial pressure in the test compartment to be determined.

As mentioned above, the major application of the oxygen sensor is in the automotive industry and literally hundreds of thousands are produced each year for this purpose to control exhaust emissions and for conserving fuel. A typical oxygen sensor based on this principle for use in car exhausts is shown in Fig. 8.2. The sensor consists of a ZrO_2/CaO (or Y_2O_3) ceramic thimble coated with porous platinum electrodes. The platinum electrodes are usually protected with a porous coating of spinel or alumina to reduce poisoning from other components of the exhaust gas. The sensor is inserted in the car exhaust pipe so that the outside of the thimble is in contact with the high temperature exhaust gases. The inside of the thimble is in contact with ambient air (p_{O_2} of 0.2 atm) which acts as the reference side of the Galvanic cell. The cell potential gives a measure of the amount of oxygen in the exhaust gas stream. Such oxygen sensors respond in fractions of a second, are simple in design, and the signal is easily digitized for transmittance to a computer.

The oxygen sensor is used in the car engine control system to adjust the ratio of air to fuel in order to minimize the nitrogen oxides (NO_x) content of the exhaust gas and to save fuel. The most appropriate air to fuel ratio (i.e. mass of air to mass of fuel) for the operation of a car engine is 14.5 since this produces amounts of NO_x, CO, and hydrocarbons which are most efficiently converted to N_2, CO_2, and H_2O in the catalytic converter. This air to fuel ratio is termed a 'rich' mixture and results in low amounts of oxygen in the exhaust gas (p_{O_2} of 10^{-15} to 10^{-46} atm). The above oxygen sensor is very sensitive under these conditions.

8.3 The diffusion-controlled limiting current oxygen sensor

Car engines are sometimes operated in a fuel 'lean' mode in order to further conserve fuel. In this mode the ratio of air to fuel is about 16 and the exhaust gas contains much higher amounts of oxygen (p_{O_2} of 10^{-2} atm). The Galvanic cell oxygen sensor described above is much less sensitive under these conditions and an alternative arrangement is normally used. This is called the 'diffusion-controlled limiting current' sensor which is also known as the

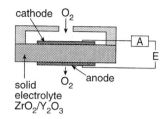

Fig. 8.3 Diffusion-controlled limiting current sensor.

'polarographic' or 'amperometric' sensor. In this, instead of simply measuring the Galvanic cell potential, a potential is applied across the platinum electrodes to 'pump' oxygen across the solid electrolyte and the electrical current flowing is measured. The construction of the sensor is illustrated in Fig. 8.3. In this configuration, the ceramic ZrO_2/Y_2O_3 (or CaO) solid electrolyte is coated with platinum electrodes and a potential is applied across the electrodes. The cathode is fully enclosed except for a small hole to let air in. The reaction at the cathode is the same as that represented by eqn 8.2 with the formation of the oxygen ion which is transported across the electrolyte to the anode side. At the anode, the oxygen ion is oxidized back to molecular oxygen according to eqn 8.3.

The gas sensing mechanism is similar in some ways to the polarographic measurement in aqueous solutions except that the reducible species (O_2) is in the gas phase. In conventional aqueous polarography (a term which strictly should only be used when the dropping mercury electrode acts as the cathode), the reducible species diffuses to the cathode surface where it is reduced under the applied potential immediately it reaches there. The measurement is said to be made under diffusion control and the current which flows is called the diffusion current or limiting current.

In the oxygen sensor, the rate-determining step is the diffusion of O_2 through the small hole. When a potential is applied across the electrodes, oxygen is transported out of the cell compartment through the solid electrolyte. More oxygen diffuses through the small hole from the outside (i.e. from the sample gas) to compensate for this. When the applied potential is increased, the rate of transport of oxygen across the electrolyte is increased and so the rate of diffusion of oxygen from the outside through the small hole increases. Eventually, the rate of oxygen transport equals the rate of diffusion into the compartment and the steady state is reached (diffusion limited region) for that particular oxygen concentration. The current is measured at the potential under which the system is diffusion limited, and is proportional to the amount of oxygen in the gas stream. Of course, the system needs to be calibrated against known oxygen concentrations. This sensor is much more sensitive than the Galvanic cell sensor at higher oxygen levels.

8.4   Other applications of the oxygen sensor

In the above discussion it has been emphasized that the major use for the solid electrolyte oxygen sensor is in the automobile, however, the sensor has other uses. For example, it has been used to determine the total oxygen demand in natural waters. In Chapter 5, the use of the oxygen probe is described for the determination of the concentration of dissolved oxygen in waters in studies of the biological oxygen demand (BOD). Oxygen in polluted waters is consumed by the decay of organic material. Another way of determining the extent of pollution of water by organic material is to measure the total oxygen demand using the solid electrolyte oxygen sensor.

This is done by evaporating a polluted water sample and heating the residue to a high temperature in the presence of a catalyst and excess oxygen. The carbonaceous material is combusted and consumes oxygen in the process. The oxygen sensor is used to determine the amount of oxygen consumed and this is related to the amount of organic material in the original water sample.

A third application of the solid electrolyte oxygen sensor is in metallurgy, particularly in the steel making industry. It is important to precisely control the oxygen content of steel products and this is done by monitoring the oxygen content in the molten steel before it is poured. The same is true in the production of nonferrous metals such as copper. Such determinations are very demanding because of the high temperatures involved, however, the ceramic-based oxygen sensor is ideal because of the refractory nature of the electrolyte. A number of designs for commercial oxygen sensors for use in molten steel have been used, some of which are single use sensors which are discarded after one measurement. Others can be used continuously such as the 'needle sensor' depicted in Fig. 8.4. This sensor, which works on the Galvanic cell principle, is made by plasma spray techniques which deposit, firstly, a layer of a Cr/Cr_2O_3 on a molybdenum metal needle followed by the ZrO_2/MgO ceramic layer. The Cr/Cr_2O_3 layer acts as one electrode in the cell and the second electrode is the molten metal itself in contact with the surface of the solid electrolyte. The electrical circuit of the cell is completed by simply using a metallic iron or molybdenum rod which is slow to melt in molten steel.

In this Galvanic cell configuration, the reference electrode is the Cr/Cr_2O_3 layer (instead of air) and, in such a metal/metal oxide layer, the oxygen pressure is fixed by the equilibrium shown in eqn 8.6 which produces a constant reference potential.

$$2Cr_{(s)} + 3/2O_{2(g)} = Cr_2O_3 \qquad (8.6)$$

The Galvanic cell potential is given by eqn 8.7.

$$E = RT/6F \ln p_{O_2} \text{ (in the molten steel)}/p_{O_2} \text{(in the reference layer)} \qquad (8.7)$$

8.5 Other solid electrolyte sensors

The solid electrolyte oxygen sensors based on ZrO_2 ceramic materials have been shown to be sensitive to other gases as well, such as H_2, N_2O, and water vapour. The response to molecular hydrogen is thought to occur by the transport of hydrogen ions through the electrolyte whereas the response to nitrous oxide occurs by an indirect mechanism involving the decomposition of the gas at high temperatures to give molecular oxygen.

There is considerable interest in developing a sensor for the oxides of sulfur (SO_2 and SO_3) because of the effects of these gases on the environment. The gases are released into the air through the burning of sulfur containing fossil fuels and give rise to acid rain. One application of this sensor would be to monitor the stack gases from coal burning power stations. A successful sensor for the oxides of sulfur needs to transport the sulfate ion (SO_4^{2-}) and a number of materials have been investigated for this including molten alkali metal sulfate mixtures. However, none have been successful enough to date to make them suitable for commercial application.

8.6 Semiconductor gas sensors

A group of sensors which can be used for the detection of gases is based on the semiconductor properties of certain metal oxides, particularly transition metal oxides, and these sensor types have been studied extensively. The first

Oxygen which is present in molten metals forms oxides on solidification of the metal which can cause cracking when the metal is subjected to stress.

Over half a million oxygen sensors are used in the steel making industry each year in Japan alone.

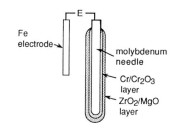

Fig. 8.4 Oxygen sensor for use in molten steel.

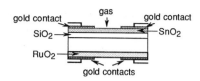

Fig. 8.5   SnO$_2$-based thin film gas sensor.

report of such a sensor was in 1962 when it was shown that a thin film, zinc oxide sensor operating at 400 °C could be used to detect CO$_2$ and organic vapours of toluene, benzene, diethyl ether, ethyl alcohol, and propane down to 1–50 ppm. Zinc oxide is an example of a semiconductor whose electrical conductivity is determined by the adsorption of species on its surface and it is the measurement of the conductivity of the metal oxide which is used to sense the presence of a gas.

Another oxide with similar semiconductor properties to ZnO is SnO$_2$ and an example of a gas sensor based on SnO$_2$ is shown in Fig. 8.5. This sensor is made by vapour deposition and other techniques used in the microelectronics industry. The sensing layer of palladium doped SnO$_2$, ~0.3 microns thick, is grown on a SiO$_2$ layer on a ferrite substrate. A thick film of RuO$_2$ is deposited on the other face of the ferrite and acts as a heater to provide the temperatures for semiconduction. Gold electrical contacts are printed on the SnO$_2$ and RuO$_2$ surfaces and the electrical resistance of the SnO$_2$ film is measured between its two gold contacts. This sensor was designed to detect ethanol and carbon monoxide.

Both p-type and n-type semiconductors have been used in sensors but the latter are preferred because they give an electrical resistance change from high to low in the presence of the gas being sensed (for a p-type semiconductor, the change in resistance is the reverse). The crystal structure of an n-type semiconductor such as SnO$_2$ contains a small excess of electrons and on exposure of the sensor to atmospheric air, oxygen is chemisorbed on the surface with each oxygen molecule consuming two electrons as given by eqn 8.8.

$$O_2 + 2e^- \rightarrow 2O^-_{ads} \tag{8.8}$$

This reaction of the surface with oxygen leads to a low electrical conductivity which is measured as a high electrical resistance. On contact with the gas being sensed, the surface reaction shown in eqn 8.9 occurs. In this, the gas removes the chemisorbed oxygen and, in doing so, is oxidized. Thus, gases which can be oxidized, i.e. which act as reducing agents can be sensed with this sensor type.

$$G + O^-_{ads} \rightarrow GO_{des} + e^- \tag{8.9}$$

This reaction produces an electron which is seen as an increase in the electrical conductivity of the SnO$_2$ layer, i.e. a decrease in the electrical resistance which is proportional to the amount of gas. In a sense, the gas can be considered as an electron donor.

ZnO is also an n-type semiconductor and operates by a mechanism similar to SnO$_2$. One problem with semiconductor gas sensors is their lack of selectivity since any reducing gas can lead to a response and considerable research is being carried out to overcome this deficiency. This is being achieved by varying the operating temperature and the addition of other elements, such as the noble metals platinum and palladium, to the surface film in order to catalyse reactions with particular gases. Also, the use of gas filters is being investigated to prevent interfering gases from reaching the sensor surface.

8.7 Applications of semiconductor gas sensors

Gas sensors similar to that shown in Fig. 8.5 can be mass produced very cheaply, by a combination of thin and thick film techniques used in the modern microelectronics industry, and are commercially available. They are used extensively as flammable gas sensors in buildings and homes.

Also, as expected, one of the largest demands for semiconductor gas sensors is in the automobile industry since they are small, simple, and cheaper to produce than solid electrolyte sensors. In this application, the sensors are used to monitor the oxygen content of the exhaust gases. The electrical resistance of the active layer (TiO_2 which is also an n-type semiconductor) increases with increase in the amount of oxygen.

There has been considerable work done on the use of semiconductor sensors for humidity detection. It has been shown that the electrical resistance of semiconductor metal oxide surfaces decreases on the adsorption of water. This effect can be used to measure humidity in the air, however, the mechanism by which the humidity sensor responds is still a matter for conjecture. A molecule like water is too stable to act as a donor to inject electrons into the valence or conduction band of the semiconductor. One suggestion is that the polar water molecule has an influence on the reaction involving the formation of the oxygen ion by adsorption of molecular oxygen (eqn 8.8) leading to a net loss in density of adsorbed O_{ads}^- ions, hence lowering the electrical resistance of the semiconductor layer.

Thin film techniques in microelectronics consist of growing layers of material on substrates using vaporization or sputtering techniques under high vacuum. Vaporization techniques are used to deposit metals by heating the metal until it vaporizes. Resistive heating is used or a high-intensity electron beam. Sputtering, however, is the preferred technique since it can be used to deposit alloys and better adhesion to the substrate is obtained. In this technique, the material to be deposited is bombarded with positive argon ions in an argon plasma. This sputters material away from the surface and it deposits on the substrate. Thick film techniques make use of screen-printing to deposit layers of material in a particular pattern on a substrate. The pattern is produced on the screen using photolithography.

8.8 Gas sensors based on MOSFETs

There are other devices from the microelectronics industry which have been investigated for their use in sensing. One of these is the metal oxide semiconductor field effect transistor (MOSFET). It has already been seen in Chapter 3 how an adaptation of this device has been used in the ion-selective field effect transistor (ISFET). The MOSFET (Fig. 8.6) is used in microelectronics, particularly in computers, for fast switching and consists of a p-type silicon wafer with two n-type regions formed by phosphorus doping. The two n-type regions are called the source and drain and a silica layer is deposited on the surface region between them. A metal layer is deposited on the silica layer and is termed the metal gate.

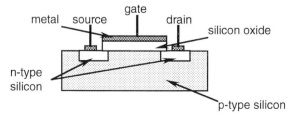

Fig. 8.6 A metal oxide semiconductor field effect transistor (MOSFET).

In the MOSFET, the application of a positive potential to the gate attracts negatively charged doping atoms to the region between the source and drain. When the applied potential reaches a certain value called the 'threshold' potential, current begins to flow between the source and drain and the amount of current is controlled by the electrical potential applied to the gate electrode.

Certain gases can be adsorbed on the gate region of the MOSFET and this process has the effect of changing the threshold potential required for current flow. The threshold potential changes according to the amount of adsorbed gas. This is the principle used in MOSFET-based gas sensors and one such sensor can detect hydrogen gas. For this, palladium metal is used as the gate metal since it can adsorb large amounts of hydrogen. The hydrogen sensor has found use in monitoring the atmosphere for hydrogen gas in industrial plants and for monitoring battery recharging. Other MOSFETs have been produced to detect gases such as H_2S and NH_3.

It seems reasonable that much progress will be made in the future on this kind of sensor since the surface reactions are used to directly actuate the devices and they can be mass produced by microelectronics fabrication techniques together with the required integrated electronic circuitry.

9 Mass sensors

9.1 The piezoelectric effect

When a chemical reaction takes place at the surface of a sensor not only is there heat evolved but there is also a change in mass. This mass change is obviously very small but if a sensitive microbalance is used, it can be detected and related to the amount of analyte reacting with the surface. Thus, there is a transducer for a chemical sensor which is based on the mass change. The quartz microbalance is one example of an extremely sensitive detector of mass changes and the principle used in the microbalance is based on the piezoelectric effect. This effect has been used more generally in mass sensitive sensors for the detection of a number of analytes.

Pierre and Jacques Curie are credited with discovering piezoelectricity in 1880 and the phenomenon involves crystals such as α-quartz which do not have a centre of symmetry. When stress such as pressure is applied to such crystals the crystal lattice is deformed and an electrical potential is developed between deformed surfaces. If electrodes are applied to the faces of a thin wafer or plate of the crystal, current will flow in an external circuit. On release of the stress a transient current will flow in the opposite direction. This effect is used as the basis for the piezoelectric chemical sensor or mass sensor. These sensors are often referred to as 'bulk wave' sensors or in their more sophisticated form as 'surface acoustic wave (SAW)' sensors.

In the actual operation of the piezoelectric sensor, the application of an alternating potential difference to the crystal causes mechanical oscillations in the crystal at the natural resonant frequency. The resonant frequency is dependent on the physical dimensions of the crystal and will alter if foreign material is deposited on the surface. Plates or wafers used in piezoelectric sensors are usually discs of diameter 1 to 1.5 cm or squares with a thickness of 0.2 mm. Metal electrodes (e.g. gold) with leads attached are deposited on the surfaces as shown in Fig. 9.1. A quartz disc of the type shown oscillates at a frequency of about 9 MHz. The frequency is measured with a frequency meter and an appropriate electronic circuit.

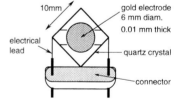

Fig. 9.1 A piezoelectric quartz crystal.

The deposition of a metal film (or any foreign substance) on the surface of the quartz causes a change in the frequency of oscillation of the quartz crystal and it is the Sauerbrey equation (eqn 9.1) which relates the mass of material deposited on the quartz crystal surface to the change in frequency. This frequency change is equivalent to the frequency change which a layer of quartz of the same mass would cause.

$$\Delta F = -2.3 \times 10^6 F^2 (M/A) \qquad (9.1)$$

ΔF = the change in frequency (Hz); F = the oscillating frequency of the quartz crystal; M = the mass of deposited material; A = the area coated by the film.

This method of determining mass by measuring the change in oscillation frequency of a quartz crystal is extremely sensitive since the type of crystal shown in Fig. 9.1 has a sensitivity of about 10^{-9} g/Hz with a detection limit of around 10^{-12} g, and frequency can be measured with high accuracy and precision.

Piezoelectric quartz crystals have found applications in chemical sensing particularly for gas and vapour sensing. A coating is deposited on the quartz crystal and, depending on the thickness of this coating, an analyte gas is adsorbed or absorbed onto or into the surface coating. Coatings can be applied by spraying, dipping, painting, or by deposition from the vapour state. The resultant change in frequency of oscillation of the quartz is related to the amount of gas adsorbed. The main difficulty with piezoelectric quartz crystal sensors for gas sensing is the lack of selectivity since nobody has succeeded in finding a coating that will adsorb only a single analyte gas. Some examples of piezoelectric chemical sensors for gases and vapours are described below.

9.2   Water vapour sensor

There is obviously considerable interest in monitoring the humidity of our environment as well as determining the amount of water vapour in such products as industrial gases. Piezoelectric sensors have been developed for these problems and such devices require a quartz crystal coated with a film of material which is hygroscopic. Many substrates have been investigated such as polymers, gelatin, silica gel, molecular sieves, and some metal fluorides. All these materials absorb water to some extent and the resultant mass change of the crystal as determined by the change in oscillation frequency is related to the concentration of water vapour in the gas or air sample. A sensor which is used as a monitor must be reversible and the water vapour needs to be removable again with minimal hysteresis by heating.

Commercial water vapour monitoring systems are available which are based on a piezoelectric sensor and the water vapour sensor seems to be the most successful of this type of chemical sensor at the present time.

9.3   Sulfur dioxide sensor

As mentioned in Section 8.5, sulfur dioxide is one gas which has been introduced into our environment by burning fossil fuels such as oil and coal and from motor vehicle engine exhausts. It is a particularly bad problem and results in acid rain in addition to causing respiratory difficulties in people living in cities. Thus, there is a need to monitor the air we breathe in an attempt to control sulfur dioxide pollution. Of course, it is not the only source of air pollution, oxides of nitrogen are equally bad and the techniques described here for sulfur dioxide could possibly be used for the oxides of nitrogen also (with the appropriate reagents).

A large number of coating materials have been investigated for sulfur dioxide detection using a piezoelectric sensor. As expected, a number of these are basic reagents such as organic amines which react with sulfur dioxide in the presence of moisture to form a salt as shown in eqn 9.2.

$$R_3N + SO_2 + H_2O \rightleftharpoons R_3NH^+HSO_3^- \tag{9.2}$$

The reaction shown in eqn 9.2 will lead to an increase in mass on the sensor surface which will be detected as a change in oscillator frequency. The sensor, of course, needs to be calibrated against known concentrations of sulfur dioxide in air.

Organic amines would not be expected to be particularly selective for sulfur dioxide, and oxides of nitrogen, for example, would interfere. One amine which has received particular attention for use in the sulfur dioxide sensor is quadrol (N,N,N',N'-tetrabis(2-hydroxypropyl)-ethylenediamine) which gives a linear behaviour for the Sauerbrey equation over the concentration range of 10 parts per billion to 30 parts per million.

9.4 Ammonia sensor

Ammonia is a pungent-smelling gas which often emanates from waste waters which are alkaline or from decaying organic wastes which contain nitrogen compounds. A number of coatings for the piezoelectric crystal have been investigated to detect ammonia in air such as ascorbic acid, L-glutamic acid hydrochloride, and pyridoxine hydrochloride. Pyridoxine hydrochloride (vitamin B_6 hydrochloride) has the ability to reversibly bind ammonia according to eqn 9.3 and the reaction leads to a mass increase which is sensed by the change in frequency of the quartz crystal. This reagent has high sensitivity (μg/L) towards ammonia but does react to other basic analytes like amines.

$$(9.3)$$

9.5 Sensor for hydrocarbons

There are many high boiling point organic liquids which are used as stationary phases in gas chromatography (e.g. Carbowax 550) and these have a certain selectivity for organic molecules depending on the solubility of the particular molecule in the stationary phase. It is not surprising then that such stationary phases have been studied as possible coatings for piezoelectric sensors. Once again, high selectivity is not a feature of such coatings but, nevertheless, some preference has been observed for aromatic hydrocarbons. In these systems, the hydrocarbon vapour dissolves in the coating thus increasing its mass.

A portable sensor system has been produced for monitoring toluene in a printing works. The sensor required a 40s response time and a 30s recovery time. Interference was observed from benzene and alkylbenzenes like *p*-xylene, ethylbenzene, and mesitylene. The system, however, compared well with two reference methods.

There are many stationary phases used in GC, they are selected depending on the type of compounds to be separated. The Carbowaxes are particularly good for separating hydrocarbons and have the general formula, $HO–(CH_2CH_2O)_x–H$.

9.6 Hydrogen sulfide sensor

Hydrogen sulfide is an extremely toxic gas and, in low concentrations, can be detected by its foul odour. However, at concentrations which are dangerous it cannot be detected by smell. Thus, a sensor for hydrogen sulfide is important

The reaction between silver acetate and hydrogen sulfide is the following.

$2Ag^+(CH_3COO^-) + H_2S \rightarrow Ag_2S + 2CH_3COOH$

to industrial health and safety. Coatings consisting of silver, copper, or lead acetates have been used to detect hydrogen sulfide and function by the reaction of the acetate salts with the gas to form the metal sulfides. The increase in mass is, of course, equivalent to the mass of hydrogen sulfide absorbed by the coating.

A highly selective sensor for hydrogen sulfide has been obtained by using a coating for the quartz crystal which consists of an acetone extract of the soot obtained by burning chlorobenzoic acid. This sensor detects hydrogen sulfide down to 1 part per million in air. The mechanism for the way in which the coating senses hydrogen sulfide is not clear, but carbonaceous material has a high affinity for adsorbing polar gases.

9.7   Mercury sensor

Mercury is the only metallic element which is a liquid in its standard state. It has an appreciable vapour pressure at ambient temperatures and pressures, and is extremely toxic. A gold-plated quartz crystal has been used to sense mercury vapour. Gold and mercury form an amalgam and this makes a convenient way to detect mercury since the mass of the gold coating will increase in the presence of mercury vapour. The sensor is not particularly easy to reverse and the gold surface must be regenerated by heating the sensor at 150 °C.

9.8   Carbon monoxide sensor

Carbon monoxide is another dangerous gas which cannot be detected by smell. Thus, considerable work has been carried out to develop sensors and monitors for carbon monoxide at the parts per billion and parts per million levels. A piezoelectric sensor for carbon monoxide has been made by using the reducing properties of the gas. The sample gas containing carbon monoxide is first reacted with mercury(II) oxide at 210 °C which produces mercury vapour according to eqn 9.4. The quantity of mercury vapour produced is related to the concentration of carbon monoxide in the gas sample and the mercury is sensed by the piezoelectric mercury sensor described in Section 9.7.

$$HgO + CO \rightarrow Hg + CO_2 \tag{9.4}$$

9.9   Sensor for explosives

There is increasing interest in developing portable chemical sensors for detecting traces of explosives in air in production plants and in public places like airports. A reasonably selective and sensitive piezoeletric sensor has been made to detect mononitrotoluenes. The coating applied to the quartz crystal in this case is the gas chromatography stationary phase, Carbowax 1000, and the organic vapour dissolves in the coating increasing its mass. Some perfumes were found to interfere.

9.10 Sensor for organophosphorus compounds

Many pesticides, particularly insecticides, are based on organophosphorus compounds, and these are cholinesterase inhibitors in animals and humans and so are extremely toxic. Thus, there is interest in developing chemical sensors to detect them in the environment. These chemicals have a phosphoryl-oxygen atom which can donate a pair of electrons to a transition metal atom and so form a coordinate bond (e.g. $-P{=}O{:}{\rightarrow}M^{n+}$). Use has been made of this to produce piezoelectric chemical sensors for organophosphorus compounds. Salts such as $FeCl_3$, $CuCl_2$, and $NiCl_2$ have been applied to the surface of a quartz crystal and, in the presence of an organophosphorus compound, react to form a complex. The use of complexes of transition metals as coatings has also been reported. In these cases, the organophosphorus compound forms a stronger complex with the metal ion than the ligand initially present.

The complexation reaction between the transition metal ion and the organophosphorus compound produces a change in mass on the surface which is measured and related to the amount of organophosphorus compound. Such coatings have also been used to detect chemical warfare agents which are similar in structure to the organophosphorus insecticides.

Organophosphorus pesticides consist of esters of phosphoric, phosphonic, and phosphinic acids. Nerve gases used in chemical warfare have similar structures.

9.11 Surface acoustic wave or SAW sensors

A more recent development in the piezoelectric sensor field involves a sophisticated device called the surface acoustic wave sensor or SAW sensor. This is an area where applied physicists and chemists have worked together in producing sensors which have considerable promise. The principle is similar to the bulk wave devices discussed above but this is extended by making use of the property of surfaces to conduct acoustic waves.

An example of a SAW sensor is shown in Fig. 9.2 in which two sets of metal electrodes are deposited onto the surface of a piezoelectric solid. One set of interleaved electrodes makes up what is called the transmitter and the other the receiver. These simply make up oscillator circuits. In between the two sets of electrodes is a region called the delay line and a chemically sensitive layer is deposited on this region. An alternating voltage is applied between the electrodes in the transmitter circuit which generates a piezoelectric acoustic wave, and this travels along the surface of the piezoelectric solid and is picked up as an electrical signal in the receiving set of electrodes.

The frequency of the generated piezoelectric wave is determined by the spacing between individual metal electrodes, and the resonant frequency of the overall device is determined by this and by the velocity of the surface acoustic wave which is propagated between the transmitter and the receiver. Sensing is achieved by the interaction of the sensing layer with the propagated wave. Any change in the surface layer mass such as occurs in chemical sensing will change the velocity of the surface acoustic wave and this will be seen as a resonant frequency change. It should be noted that the delay line does not even need to be made of piezoelectric material which provides great flexibility in the design of SAW sensor devices.

One great advantage of SAW sensors is that they can be made using microelectronic processing techniques which means the size of the metal

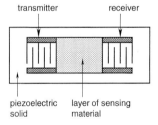

Fig. 9.2 A surface acoustic wave (SAW) sensor.

electrodes and the spacing between them can be extremely small. This allows high resonant frequencies to be produced (>1 GHz) and sensitivity to mass changes increases with increase in frequency. It has been suggested that mass changes in the femtogram (10^{-15} g) region are possible.

Because of their mode of operation, it is not unexpected that SAW sensors have been made using the types of reagents described above for bulk wave sensors for the detection of gases. There has also been considerable work carried out on 'array SAW sensors'. This is an attractive approach to sensor technology since it allows the use of reagents which show some selectivity for a particular gas but need not be highly selective. In the array sensor, a number of identical SAW sensors are grouped together but each with a different reagent coating and the device produces what is referred to as a 'fingerprint' of the gas. Then, using pattern recognition methods, and a library of array responses it is possible to identify the gas or a combination of gases. Such an approach is useful for continuous monitoring in industrial plants where the only information needed is whether or not there has been a change in the composition of a vapour stream, and it is necessary to sound an alarm.

9.12   The use of piezoelectric sensors in solution

All of the applications for bulk wave and SAW piezoelectric sensors described above involved detection of an analyte in the gas phase. However, many of the applications for chemical sensors involve sensing and determining species in solution. Chemical sensors find particular application in automated flow analysis systems like flow injection, segmented flow, and continuous flow analysis. It was initially thought that attempts to use piezoelectric-based sensors would fail because it was feared that a liquid phase in contact with the quartz crystal would damp the acoustic wave at the surface.

Recent research, however, has suggested that such sensors can be used in solutions and may be particularly useful in immunochemical procedures such as, for example, the reaction of an antigen immobilized on the surface of a quartz crystal with an antibody. Such sensors would be expected to be very sensitive because of their response to small mass changes.

9.13   Piezoelectric-based immunosensors

One example of a successful biosensor which uses a piezoelectric transducer is able to detect *Candida albicans* which is a yeast-like fungus found in humans. In this sensor, the silver electrodes of the piezoelectric crystal are firstly plated with palladium metal which is then oxidized. The anti-*Candida* antibody is immobilised onto the surface and on dipping the sensor into a suspension of *Candida*, the frequency of oscillation of the crystal changes due to the immunochemical reaction between the antibody and the fungus which causes a change in mass on the surface.

Another example of an immunochemical biosensor uses Protein A immobilized on the surface of the electrodes of the crystal to detect immunoglobins. The affinity reaction between Protein A and human IgG produces a mass change on the surface which also produces a change in the frequency.

10 The future

In this primer we have explored the fascinating world of chemical sensors and have discovered the various types of sensors which are known and the basic chemical principles involved in the way they function. Numerous chemical sensors of all types have been proposed but many of these are still scientific curiosities, particularly when it comes to biosensors, and their commercial development is yet to be achieved. Nevertheless, there are great commercial rewards in chemical sensor production and the industry is anticipated to be worth billions of dollars by the turn of the century. Chemical sensors have changed the way we think about analytical chemistry methods and clinical testing procedures, and have given us the possibility of having a hand-held or pocket analytical chemistry laboratory. The applications of chemical sensors are almost limitless in our need and desire to monitor everything around us.

There is a very new and exciting field of research called 'Miniaturized Total Analysis Systems (μTAS)' and this promises to produce complete analytical chemistry systems on a micro-scale which, a few years ago, were hard even to imagine. These systems are based on micromachining and microfabrication techniques.

The commercial production of chemical sensors is driven by this need for cheap and reliable sensors and, in order to achieve this, mass-production techniques must be used. This is very evident in the automobile industry which produces millions of cheap oxygen sensors each year and is arguably the best example of how mass-production techniques have been employed successfully to produce chemical sensors.

The mass production of chemical sensors is being achieved using techniques which are already well established in the microelectronics industry. We are all aware of the impact the 'microchip' has had on our society and how its mass production has led to extremely cheap computers and electronic equipment. Chemical sensors can also be produced using these mass-production techniques although such techniques are more compatible with certain types of sensors than others, e.g. electrochemical sensors.

Chemical sensors made with microelectronics techniques are called 'planar' sensors and can be produced using thick film techniques or thin film or integrated silicon wafer techniques. It is also possible to integrate the electronics associated with the transducer into the total package and to produce intelligent sensors or 'smart' sensors. Another advantage of these techniques is the fact that very small sensors can be produced, e.g. for *in-vivo* use, it may be necessary to incorporate a sensor or several sensors into the tip of a catheter or needle. Other devices may make use of an array of dozens of chemical sensors linked together. Example of these applications can be found in the sensor literature.

An 'electronic nose' consisting of 15 sensors has been used to test the gases emanating from tomatoes in a study of the effects of radiation on vegetables. A similar 'nose' has been used to determine the age of cod fillets.

Considerable work has been carried out on the production of thick film sensors for electrochemical measurements and these are manufactured by screen-printing layers onto a substrate which is often a flat ceramic or alumina plate. The process is not unlike conventional printing and the layers are printed as thick pastes by pressing the paste through a pattern on a screen. The pattern on the screen is made by photolithography. The various layers on the substrate are then dried or fired. Layers can consist of metals to produce electrically conducting pathways and polymer membranes to produce sensors.

electrical contacts
plug directly into →
pen-size or pocket
meter

sensing + reference
← electrodes on which a
drop of blood is placed

Fig. 10.1　Disposable blood-glucose sensor strip.

The glucose biosensor described in Section 5.4 is made in this way and is one of the first examples of a thick film sensor which has been produced commercially. The disposable sensor strip which is shown in Fig. 10.1 (40 mm × 5 mm) plugs into a pen-size or pocket meter. A drop of blood is deposited on the biosensor region and the glucose concentration is determined amperometrically as described before. The substrate for the sensor is a plastic strip on which are printed a silver-based paste for the reference electrode and a carbon-based paste for the working electrode. Glucose oxidase and a ferrocene derivative are incorporated onto the graphite electrode and the surface is protected with a special coating. There is much activity in this area and it can be expected that numerous chemical sensors will be manufactured commercially in this way in the future.

Integrated silicon wafer techniques are being used to produce 'smart' electrochemical sensors which have, in addition to the chemical sensor, all the electronics necessary to process the signal. Such devices can be extremely small and can incorporate a number of sensors on the one chip. An example, shown in Fig. 10.2, is an eight-channel sensor chip which has been designed for potentiometric sensors. On the chip, which measures 3.3 mm × 2.8 mm, are eight different sensor pads with their potential buffering circuits and a reference electrode. In some ways the individual sensors are like ISFETs except that there is an extended metal contact from the FET to the sensor pad. This type of device is sometimes termed an EGFET (extended gate field effect transistor). Individual sensor membranes can be deposited on the sensor pads to produce an eight-function 'smart' sensor for the simultaneous determination of a range of cations and anions.

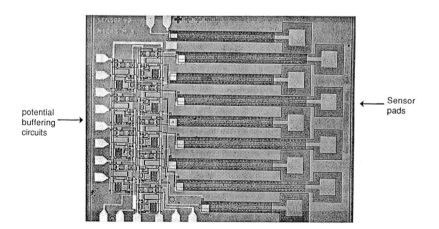

potential
buffering →
circuits

Sensor
← pads

Fig. 10.2　Eight-potentiometric sensor 'smart chip' (dimensions, 3.3 mm × 2.8 mm).
(Reproduced with permission of Institute of Physics Publishing.)

Another recent report describes a complete miniature electrochemical system in which the electrode assembly, a potentiostat, and a voltage ramp generator have all been fabricated in the one unit using thick film and hybrid techniques to produce a portable instrument for water analysis.

These are just a few examples of the technology that is being employed to produce the sensing systems that we will be using in the future. There is little doubt that the portable, miniature, and intelligent sensing devices which were mentioned at the start of this primer will become a reality in the not too distant future, in fact, it is happening already.

Further reading

Books

Janata, J. (1989). Principles of chemical sensors. Plenum, New York.

Edmonds, T. E. (ed.) (1988). Chemical Sensors. Chapman and Hall, New York.

Morf, W. E. (1981). The principles of ion-selective electrodes and of membrane transport. Akademiai Kiado, Budapest.

Covington, A. K. (ed.) (1979). Selective electrode methodology, Vols I and II. CRC Press, Boca Raton.

Lakshminarayanaiah, N. (1979). Membrane electrodes. Academic Press, New York.

Guilbault, G. G. (1984). Analytical uses of immobilized enzymes. Marcel Dekker, New York

Gopel, W., Hesse, J., and Zemel, J. N., (eds) (1991). Sensors—A comprehensive survey. VCH, Weinheim, Fed. Rep. Germany.

Edelman, P. G. and Wang, J. (eds) (1992). Biosensors and chemical sensors— Optimizing performance through polymeric materials, ACS Symposium Series 487. ACS, Washington, DC.

Mathewson, P. R. and Finley, J. W. (eds) (1992). Biosensor design and application, ACS Symposium Series 511. ACS, Washington, DC.

Murray, R. W., Dessy, R. E., Heineman, W. R., Janata, J., and Seitz, W. R. (eds) (1989). Chemical sensors and microinstrumentation, ACS Symposium Series 403. ACS, Washington, DC.

Czarnik, A. W. (ed.) (1993). Fluorescent chemosensors for ion and molecular recognition, ACS Symposium Series 538. ACS, Washington, DC.

Lambrechts, M. and Sansen, W. (1992). Biosensors : Microelectrochemical devices. Institute of Physics Publishing, Bristol, UK.

Kazuhiro Sylvester Goto (1988). Solid state electrochemistry and its applications to sensors and electronic devices. Elsevier, New York.

Reviews

Moody, G. J. and Thomas, J. D. R. (1979). Progress in designing calcium ion-selective electrodes. Ion-Selective Electrode Reviews, **1**, 3.

Cattrall, R. W. and Hamilton, I. C. (1984). Coated-wire ion-selective electrodes. Ion-Selective Electrode Reviews, **6**, 125.

Seitz, W. R. (1988). Chemical sensors based on immobilized indicators and fiber optics. CRC Critical Reviews in Analytical Chemistry, **19**, 135.

Arnold, M. A. and Meyerhoff, M. E. (1988). Recent advances in the development and analytical applications of biosensing probes. CRC Critical Reviews in Analytical Chemistry, **20**, 149.

Norris, J. O. W. (1989). Current status and prospects for the use of optical fibres in chemical analysis. Analyst, **114**, 1359.

Azad, A. M., Akbar, S. A., Mhaisalkar, S. G., Birkefeld, L. D., and Goto, K. S. (1992). Solid-state gas sensors: a review. Journal of the Electrochemical Society, **139**, 3690.

Alder, J. F. and McCallum, J. J. (1983). Piezoelectric crystals for mass and chemical measurements. Analyst, **108**, 1169.

McCallum, J. J. (1989). Piezoelectric devices for mass and chemical measurements. Analyst, **114**, 1173.

Index